ADVANCED PROCESS CONTROL
AND SIMULATION
FOR CHEMICAL ENGINEERS

ADVANCED PROCESS CONTROL AND SIMULATION FOR CHEMICAL ENGINEERS

Hossein Ghanadzadeh Gilani, PhD,
Katia Ghanadzadeh Samper, and Reza Khodaparast Haghi

Apple Academic Press

TORONTO   NEW JERSEY

© 2013 by
Apple Academic Press Inc.
3333 Mistwell Crescent
Oakville, ON L6L 0A2
Canada

Apple Academic Press Inc.
9 Spinnaker Way, Waretown, NJ 08758
USA

First issued in paperback 2021

Exclusive worldwide distribution by CRC Press, a Taylor & Francis Group

ISBN 13: 978-1-77463-263-5 (pbk)
ISBN 13: 978-1-926895-32-1 (hbk)

Library of Congress Control Number: 2012951940

Library and Archives Canada Cataloguing in Publication

Gilani, Hossein G.

Advanced process control & simulation for chemical engineers/Hossien G. Gilani, Katia G. Samper and Reza K. Haghi.

Includes bibliographical references and index.
ISBN 978-1-926895-32-1
1. Chemical process control. 2. Chemical process control–Simulation methods. 3. Chemical engineering–Simulation methods. I. Samper, Katia G II. Haghi, Reza K III. Title.

TP155.75.G54 2013   660'.2815   C2012-906417-3

About the Authors

Hossein Ghanadzadeh Gilani, PhD

Hossein Ghanadzadeh Gilani, PhD, received his MSc in Chemical Engineering from Bologna University, Italy, in 1982, and his PhD degree in Chemical Engineering from the University of Catalunya, BarcelonaTech (UPC), Spain, in 1992. He is currently an associate professor in the Department of Chemical Engineering at the University of Guilan. He has served as the Head of the Chemical Engineering Department as well as Vice-Dean of Faculty of Engineering in Research. He is a reviewer for several international journals and a member of the editorial boards of several journals. He has published more than 48 papers in various international research journals and is currently actively engaged in research areas of separation processes using distillation and extraction liquid-liquid, membranes and adsorption, process development, cryogenics, and gas liquefaction processes.

Katia Ghanadzadeh Samper

Katia Ghanadzadeh Samper is a chemical engineer and is currently in the postgraduate program of chemical systems at the University of Barcelona, Spain.

Reza Khodaparast Haghi

Reza Khodaparast Haghi is a mechanical engineer and is currently in the postgraduate program of Advanced Control Systems at University of Salford, Manchester, (UK).

Contents

List of Abbreviations

ANOVA	Analysis of variance
ANN	Artificial neural network
AFD	Average fiber diameter
BM	Bukacek-Maddox
CCD	Central composite design
CA	Contact angle
DOE	Design of experimental
DMF	Dimethylformamide
FTIR	Fourier transforms infrared spectroscopy
GC	Gas chromatographic
GBFS	Granulated blast-furnace slag
HBPs	Hyperbranched polymers
IROST	Iranian Research Organization for Science and Technology
MLP	Multilayer Perception
NRTL	Non-random two-liquid
NMR	Nuclear magnetic resonance
OPC	Ordinary Portland cement
OA	Orthogonal array
PAN	Polyacrylonitrile
PET	Polyethylene terephthalate
PP	Polypropylene
RSM	Response surface methodology
RMSE	Root-mean-square error
SEM	Scanning electron micrographs
SI	Severity index
SSE	Squares due to error
TBA	Tert-butanol
TCD	Thermal conductivity detector
UV–vis	Ultraviolet–visible
UNIQUAC	Universal quasi-chemical
VLE	Vapor-liquid equilibrium
WPLA	Waste PET bottles lightweight aggregate
WPLAC	Waste PET bottles lightweight aggregate concrete
XRD	X-ray diffraction

List of Symbols

if = Liquid environment near the feed phase
of = Organic environment near the feed phase
os = Organic environment near the stripping phase
is = Liquid environment near the stripping phase
o = Organic phase
i = Liquid phase
δ = Thickness of Mass transfer film
D = Mass transfer diffusion coefficient
N = Mass transfer diffusion
a = optimized interaction parameter
A, B, and C = Antoine equation parameters
B_{ij} = 2^a coefficient of the virial
C = number components
Calc = calculated value
E = excess property
exp = experimental value
G_{ij} = adjustable parameter
2EH = iso octhyl alcohol
NBA = n.butanol
q_i = relative surface area per molecule
r_i = number of segments per molecule
T = absolute teperture (Kelvin)
u_{ij} = interaction energy
x = mole fraction
x_i = equilibrium mole fraction of component i
Z = lattice coordination number, set equal to 10
z_i = number of moles of component i

Greek Symbols
F = segment fraction
q = area fraction
g = activity coefficient
t_{ij} = adjustable parameter in the UNIQUAC equation

Superscript
c = combinatorial part of the activity coefficient
q = UNIQUAC equation
r = residual part of the activity coefficient
i = ith component

Preface

This book offers a modern view of process control in the context of today's technology. It provides the standard material in a coherent presentation and uses a notation that is more consistent with the research literature in process control. The purpose of this book is to convey to students an understanding of those areas of process control that all chemical engineers need to know. The presentation is concise, readable, and restricted to only essential elements. Topics that are unique include a unified approach to model representations, process model formation and process identification, multivariable control, statistical quality control, and model-based control. The methods presented have been successfully applied in industry to solve real problems. This book is designed to be used as an advanced research guide in process dynamics and control. In addition to chemical engineering courses, the book would also be suitable for mechanical, nuclear, industrial, and metallurgical engineers. The book offers scope for academics, researchers, and engineering professionals to present their research and development works that have potential for applications in several disciplines of engineering and science.

— **Hossein Ghanadzadeh Gilani, PhD,**
Katia Ghanadzadeh Samper, and Reza Khodaparast Haghi

1 Artificial Neural Network (ANN) Models and Polymers

CONTENTS

1.1 INTRODUCTION

Hyperbranched polymers (HBPs), due to their unique chemical and physical properties, have attracted increasing attention. These polymers are highly branched, polydisperse, and three-dimensional macromolecules [1]. The HBPs have remarkable properties, such as low melt and solution viscosity, low chain entanglement, and high solubility, as a result of the large amount of functional end groups and globular structure, so they are excellent candidates for use in random applications, particularly for modifying fibers [2, 3].

Recently, the application of HBPs in textile industry has been developed. For instance, in the study on applying HBP to cotton fabric [4-8], it was demonstrated that HBP treatment on cotton fabrics has no undesirable effect on mechanical properties of fabrics. Furthermore, application of HBP to cotton fabrics reduced UV transmission and has good antibacterial activities. The study on dyeability of polypropylene (PP) fibers modified by HBP showed that the incorporation of HBP prior to fiber spinning considerably improved the color strength of PP fiber with C.I. Disperse Blue 56 and has no significant effect on physical properties of the PP fibers [9]. Literature review showed that there has not been a previous report regarding the treatment of amine terminated HBPs on PET fabric and study of its dyeability with acid dyes. In the most recent study in this field, fiber grade PET was compounded with polyesteramide HBP and dyeability of resulted samples with disperse dyes was studied [10]. The results showed that the dyeability of dyed modified samples comprised of fiber grade PET films and a HBP (Hybrane H1500) were better than the neat PET and this was increased by increasing amount of HBP in presence or absence of a carrier. The dyeability of the samples was attributed to decrease in glass transition temperature for blended PET/HBP in comparison with neat PET [10].

In this study, the effect of HBP treatment parameters such as solution concentration, treatment temperature and time on dyeuptake (K/S value) of PET fabric were investigated using ANN models based on a feed forward topology.

1.2   EXPERIMENTAL

1.2.1   Material

A HBP with amine terminal group were synthesized and characterized as described by previous research [4]. Figure 1 represent the structural units of HBPs include terminal,

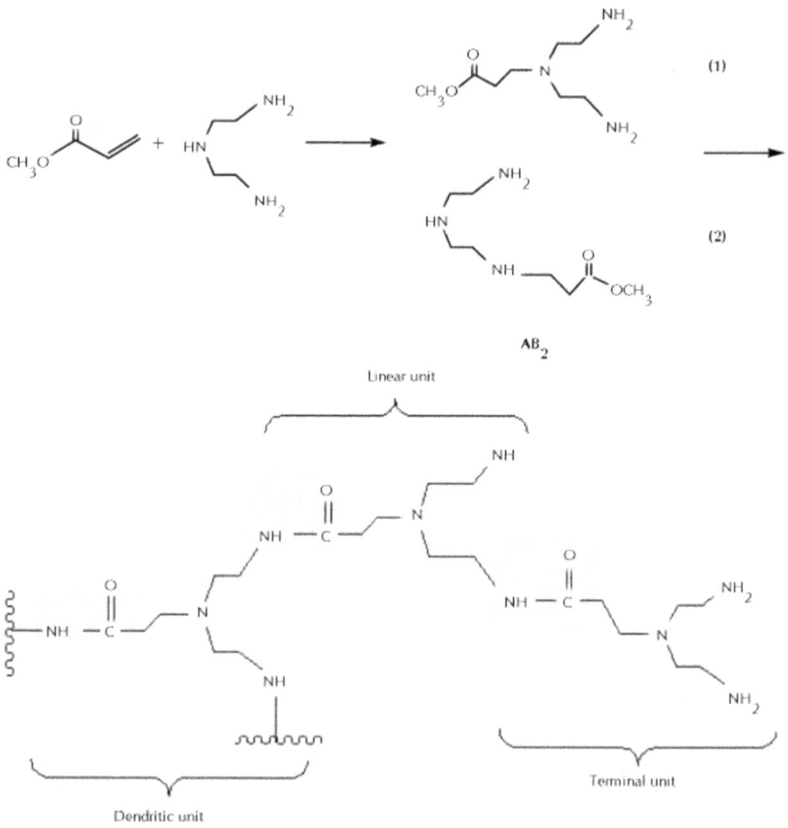

FIGURE 1   Chemical structure of amine terminated hyperbranched polymer.

dendritic and linear units. The PET fabric (28×19 count/cm^2) used throughout this work and before use it was treated with a solution containing 5 g/l Na$_2$CO$_3$ and 1 g/l of a non-ionic detergent at 60°C for 30 min to remove undesired materials. Distilled water was used for the treatments and washings. Acid dye (C.I. Acid Red 114) was provided by the Ciba Ltd. (Tehran, Iran) and used to evaluate the dye absorption behavior (dyeability) of samples.

1.2.2 PET Fabric Treatment with HBP

The PET fabric samples were immersed in aqueous solution of sodium hydroxide (10% w/v) at the temperature of 94°C for 1 hr with the liquor to goods ratio of 40:1. Then the samples were thoroughly rinsed with distilled water and neutralized with acetic acid, and finally rinsed and dried at room temperature. After the alkaline treatment of fabrics, the HBP was applied to samples using exhaustion process. The alkali-treated PET fabrics were treated with HBP solution and the temperature was raised at a rate of 2.5°C/min. After exhaustion, the samples were thoroughly rinsed with distilled water to remove unfixed HBP and dried at room temperature. Figure 2 show the HBP treatment profile.

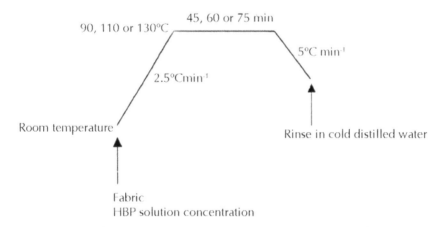

FIGURE 2 Method and graph for HBP treatment of PET fabric.

1.2.3 Dyeing Procedure

The HBP treated fabrics were introduced into the dye baths containing dye (C.I. Acid Red 114) with the liquor to goods ratio of 60:1 at the temperature 40°C, increasing to boil with the constant rate of 3°Cmin^{-1}. Dyeing was then continued for 60 min with occasional stirring. At the end of dyeing the dyed samples were rinsed with cold water, then with hot water at about 50°C and finally rinsed with tap water.

1.2.4 Measurement and Characterization

The CIELAB color coordinates of dyed samples were determined under illuminant D$_{65}$ at 10°C standard observers in the visible range using color-eye 7,000A spectrophotometer. As shown by Equation (1), the Kubelka–Munk single-constant theory was employed to calculate K/S values at the wavelength of maximum absorption (λ_{max}) for each fabric [11].

$$(K/S)_{c,\lambda} = \frac{\left(1 - R_{M,\lambda}\right)^2}{2R_{M,\lambda}} - \frac{\left(1 - R_{S,\lambda}\right)^2}{2R_{S,\lambda}} \tag{1}$$

where, $R_{M,\lambda}$ and $R_{S,\lambda}$ are the reflectance values at the wavelength of maximum absorbance (λ_{max}) for colored and uncolored substrate respectively, K is the absorption coefficient and S is the scattering coefficient. The HBP treatment conditions and samples K/S value are shown in Table 1.

TABLE 1   The HBP treatment parameters and response.

No.	Inputs			Output
	HBP concentration (wt.%)	Temperature (°C)	Time (min)	K/S value
1	2	90	75	20.38
2	2	130	75	21.57
3	2	110	60	19.86
4	2	130	45	21.09
5	2	90	45	19.78
6	6	110	60	22.68
7	6	110	60	22.18
8	6	110	60	22.67
9	6	110	60	22.97
10	6	110	60	22.64
11	6	110	60	22.26
12	6	110	45	22.59
13	6	110	75	24.47
14	6	90	60	22.58
15	6	130	60	26.07
16	10	90	45	23.35
17	10	90	75	24.97
18	10	130	75	29.92
19	10	130	45	26.01
20	10	110	60	23.23

1.2.5 Artificial Neural Network

The ANN as an information processing technique, are composed of simple unit operating in parallel. A typical ANN represents a network with a several number of layers, consisting of an input layer, one or more hidden layers and an output layer. The ANN can be trained to perform a particular function by adjusting the values of the connections (weights) between elements. The weights between the neurons play an important role during the training process. The interconnection weights are adjusted, based on a comparison of the network output and the actual output, to minimize the error between the network output and the actual values [12, 13].

In this work, the multilayer perceptron ANN was used to process data using the modified backpropagation algorithm. The ANN has three inputs referred to HBP treatment parameter (HBP solution concentrations, treatment temperature and treatment time), and one output referred to K/S value of treated samples. All calculations carried out in Matlab mathematical software (version 7.6) with ANN toolbox. The various topology of neural network are shown in Table 2.

TABLE 2 The topology of neural networks.

No. of model	Number of hidden layer	Number of neuron per hidden layer
1	1	3
2	1	4
3	1	5
4	1	6
5	1	7
6	1	8
7	2	3-2
8	2	4-2
9	2	4-3
10	2	3-3

1.3 RESULTS

In this work, the dyeability of HBP treated PET fabrics is calculated from treatment condition comprising HBP solution concentrations, treatment temperature and treatment time using ANN. The prediction performance is evaluated by calculating R^2 and RMSE through the following equation:

$$RMSE = \sqrt{\frac{\sum_{i=1}(y_{i,\exp} - y_{i,pred})^2}{n}} \qquad (2)$$

where, $y_{i,\exp}$ and $y_{i,pred}$ are the experimental and predicted values, respectively, and n is the number of the experimental run.

The suitable number of neurons in the hidden layer was determined by changing the number of neurons. As shown in Table 3, the best prediction, based on minimum error, was obtained by ANN with 8 neurons in hidden layer. The R^2 and RMSE were 0.97 and 0.61 respectively. The comparison between the actual and predicted value given by ANN is shown in the Figure 3 and demonstrated that all points are located close to a straight line.

TABLE 3   The results of prediction by artificial neural network.

No. of model	Mean	SD	Max.	Min.	R^2	RMSE
1	0.03	0.04	0.13	0.01	0.85	1.21
2	0.05	0.04	0.16	0.00	0.84	1.42
3	0.05	0.04	0.20	0.00	0.82	1.57
4	0.04	0.03	0.11	0.00	0.89	1.15
5	0.03	0.03	0.10	0.01	0.91	0.96
6	0.02	0.02	0.09	0.00	0.97	0.61
7	0.03	0.03	0.15	0.01	0.93	0.91
8	0.03	0.03	0.12	0.00	0.89	1.11
9	0.03	0.03	0.12	0.00	0.90	1.04
10	0.04	0.04	0.15	0.00	0.87	1.24

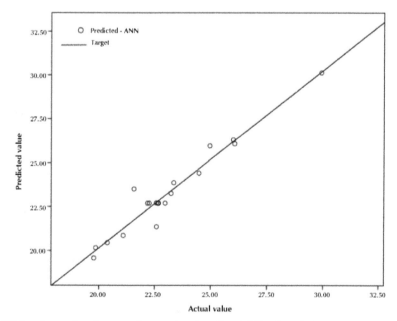

FIGURE 3   Comparison between the actual and predicted K/S value of HBP treated PET fabrics.

1.4 CONCLUSION

In this work, ANN models were developed using a feed forward topology for modeling of HBP treatment on PET fabric. The effects of three HBP treatment parameters namely solution concentrations (wt.%), treatment temperature (°C) and time (min) on dyeability (K/S value) of treated PET fabrics were investigated. The best prediction was obtained by ANN with 8 neurons in hidden layer. In this model the R^2 and RMSE were 0.97 and 0.61 respectively. Furthermore, the mean, standard division, maximum and minimum error are 0.02, 0.02, 0.9, and 0 respectively.

KEYWORDS

- **Artificial neural network**
- **Back-propagation algorithm**
- **Hyperbranched polymers**
- **Kubelka–Munk single-constant theory**
- **Polypropylene**

REFERENCES

1. Gao, C. and Yan, D. Hyperbranched polymers: From synthesis to applications. *Progress in Polymer Science*, **29**, 183–275 (2004).
2. Schmaljohann, D., Pötschke, P., Hässler, R., Voit, B. I., Froehling, P. E., Mostert, B., and Loontjens, J. A. Blends of amphiphilic, hyperbranched polyesters, and different polyolefins. *Macromolecules*, **32**, 6333–6339 (1999).
3. Seiler, M. Hyperbranched polymers: Phase behavior and new applications in the field of chemical engineering. *Fluid Phase Equilibria*, **241**, 155–174 (2006).
4. Zhang, F., Chen, Y., Lin H., and Lu, Y. Synthesis of an amino-terminated hyperbranched polymer and its application in reactive dyeing on cotton as a salt-free dyeing auxiliary. *Coloration Technology*, **123**, 351–357 (2007).
5. Zhang, F., Chen, Y., Lin H., Wang, H., and Zhao, B. HBP-NH$_2$ grafted cotton fiber: Preparation and salt-free dyeing properties. *Carbohydrate Polymer*, **74**, 250–256 (2008).
6. Zhang, F., Chen, Y. Y., Lin, H., and Zhang, D. S. Performance of cotton fabric treated with an amino-terminated hyperbranched polymer. *Fibers and Polymers*, **9**, 515–520 (2008).
7. Zhang, F., Chen, Y., Ling, H., and Zhang, D. Synthesis of HBP-HTC and its application to cotton fabric as an antimicrobial auxiliary. *Fibers and Polymers*, **10**, 141–147 (2009).
8. Zhang, F., Zhang, D., Chen, Y., and Lin, H. The antimicrobial activity of the cotton fabric grafted with an amino-terminated hyperbranched polymer. *Cellulose*, **16**, 281–288 (2009).
9. Burkinshaw, S. M., Froehling, P. E., and Mignanelli, M. The effect of hyperbranched polymers on the dyeing of polypropylene fibres. *Dyes and Pigments*, **53**, 229–235 (2002).
10. Khatibzadeh, M., Mohseni, M., and Moradian, S. Compounding fiber grade polyethylene terephthalate with a hyperbranched additive and studying its dyeability with a disperse dye. *Coloration Technology*, **126**, 269–274 (2010).
11. McDonald, R. Color physics for industry. *Society of dyers and colorists*, England (1997).
12. Dev, V. R. G., Venugopal, J. R., Senthilkumar, M., Gupta, D., and Ramakrishna, S. Prediction of Water Retention Capacity of Hydrolysed Electrospun Polyacrylonitrile Fibers Using Statistical Model and Artificial Neural Network. *Journal of Applied Polymer Science*, **113**, 3397–3404 (2009).
13. Galushkin, A. L. *Neural networks Theory*. Springer, Moscow Institute of Physics & Technology (2007).

2 New Trends and Achievements in Extraction of Copper

CONTENTS

2.1 INTRODUCTION

Membrane separation is an area deserving special attention because of its great potential for low capital cost and energy efficiency. To date, however, few membrane processes other than reverse osmosis and hydrogen separation have demonstrated any industrial utility primarily because of problems of speed and selectivity in separation [1].

Liquid membranes with impressive properties such as high selectivity and efficient consumption of energy in separation processes seem to be more suitable. Other advantages such as variety of configuration and carriers for different applications, simplicity of assembling and high rate of mass transfer are facilitated the implementation of these membranes. Chromium is one of contaminants that exist in waste water of various industries like steel, pigment, and leather tanning. In this project, separation of Cu (II) ion by implementation of a bulk liquid membrane using alamine as carrier, kerosene as solvent, sodium hydroxide as stripping product phase, dodecanol for preventing from jellying of inorganic, and organic phases have been investigated. Effective parameters on separation of Cu (II) ion including feed phase pH, stripping phase molarity, mixer rotational rate in feed and stripping phase, volume percentage of carrier in organic phase, presence/absence of surfactant in organic phase have been studied. In the range of designed experimental, the optimum conditions as follow have been found: ph = 2.45, Stripping phase molarity = 3, mixer rotational rate in both inorganic phases = 100 rpm, volumetric amount of carrier = 1%, and presence of surfactant.

2.2 BULK LIQUID MEMBRANE

This set up is useful only for laboratory experiments, and is setup as follows. Following Figure 1, a U-tube cell is used, and some type of carrier, perhaps dissolved in CH_2Cl_2, is placed in the bottom of the tube. That is the organic membrane phase.

Two aqueous phases are placed in the arms of the U-tube, floating on top of the organic membrane. With a magnetic stirrer rotating at fairly slow speeds, in the range of 100–300 rpm, the transported amounts of materials are determined by the concentrations in the receiving phase. Stability is maintained so long as the stirrer is not spinning too quickly.

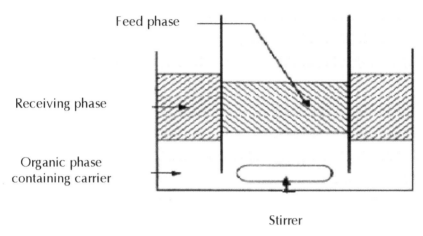

FIGURE 1   Bulk liquid membrane.

2.3   EXPERIMENTAL

For doing experiments, we used glass cell which divided to two compartments by thin sheet of glass. Dimensions of cell were 10 cm × 24 cm × 20 cm. Volumes of one section were three times greater than other section. In bottom of each section and 2 cm upper then the bottom, two holes were arranged for sampling. Both holes sealed with plastic caps. Feed phase and stripping phase were pouring in greater and little section respectively. Membrane phase was pour over foregoing both phases. Two pirex blades rotated by mixer with constant speed 15 cm × 1 cm diameter. Each one of blades has two paddles. Potassium chromate and NaoH were pouring with deter-mined concentration in grate and little section of cell respectively. Membrane phase was pour over two foregoing phases in two steps. First, Kerosene was added without carrier (for example 350 ml), then 41 ml Kerosene + 5 ml Alcohol + 4 cc Alamine was added to membrane phase. To avoid of vaporization of membrane phase top of the cell was covered with foil. Sampling was started step by step up to 24 hr (1 day). At the beginning of experiments, the color of feed phase was yellow. Membrane phase and stripping phase were color less. Gradually, the color of feed phase ap-proached to pale yellow and simultaneously the color of membrane and stripping phases approached to very pale yellow (colorless) and the color of stripping phase approached to dark yellow. This variation in the color of feed and stripping phases showed the process of extraction of chromium ion from feed phase to stripping phase. During the experiments, the pH of feed phase was controlled and sulfuric acid was added to feed phase for maintaining the pH.

2.4   PARAMETERS

Parameters were investigated as follow:

(1) PH of feed phase.
(2) Percentage of carrier in membrane phase.
(3) Presence/absence of Dodecanol in membrane phase.
(4) Molarity of stripping phase.
(5) Velocity of mixer in feed phase.
(6) Velocity of mixer in stripping phase.

Where pH was investigated in three states like: 2/4/6, Presence of dodecanol in membrane phase was investigated in three states like: 0.5, 1, and 2%. Molarity of stripping phase was investigated in three states like: 1/2/3, velocity of mixer in both liquid phases was investigated in three states like: 80, 100, and 120 rpm.

2.5   DESCRIPTION OF MODEL EXPERIMENT

2.5.1   Feed Phase

Solving 1.94 g potassium chromate (K_2CrO_4) in 2 L DM water, (0.005 mol). Adding 5 ml sulfuric acid for approaching to solution with pH = 2. Diluting 100 ml of foregoing solution to 1 L with DM water 750 ml of final solution was our feed phase.

2.5.2   Stripping Phase

Solving 30 g NaOH on 250 ml DM water.

2.6   ERROR ANALYSIS

For calculation of ions concentration error Formula (1) was used:

$$\bar{X} \pm \frac{ts}{\sqrt{N}} \tag{1}$$

where:
$\bar{X}$ = Average of three amounts of ion concentration obtained from calibration curve.
T = 95% safety limit with constant amount = 4.3
N = obtained samples
S = Standard deviation
where

$$S = \sqrt{\frac{\sum (\bar{x} - x_i)^2}{N - 1}}$$

In standard deviation formula we have $\bar{X}$ and x_i:
$\bar{X}$ and x_i are average of three amounts of ion concentration and ion concentration in each sampling respectively.

2.7   GOVERNING EQUATIONS IN TRANSPORTATION OF CHROMIUM

(a) Diffusion of chromium from bulk of feed to interface of feed phase and membrane phase:

$$N_{CrO_4^{2-}} = \frac{D_f}{\delta_{if}}\left(\left[CrO_4^{2-}\right]_f - \left[CrO_4^{2-}\right]_{if}\right)$$

(b) Reaction on the interface of membrane phase and feed phase:

$$Cro_4^{-2}(if) + 2\,AlmoH_{(of)} \leftrightarrow Alm_2Cro_4(of) + 2oH^-(if) \tag{2}$$

$$K = \frac{[Alm_2Cro_4]_{if}[OH^-]_{if}^2}{[AlmoH]_{of}[Cro_4^{-2}]_{if}} \tag{3}$$

and following reaction:

$$H^+ + OH^- \leftrightarrow H_2O \qquad K = \frac{1}{K_w} = 10^{14} \tag{4}$$

(c) Diffusion of product complex of up reaction from interface between feed phase and membrane phase to bulk of membrane:

$$N_{Alm_2Cro_4} = \frac{D_o}{\delta_{of}}\left([Alm_2Cro_4]_{of} - [Alm_2Cro_4]_o\right) \tag{5}$$

(d) Diffusion of product complex of reaction (2) from bulk of membrane to interface of membrane and stripping phase:

$$N_{Alm_2}Cro_4 = \frac{Do}{Dos}\left([Alm_2Cro_4]_o - [Alm_2Cro_4]os\right) \tag{6}$$

(e) Reaction in the interface of membrane and stripping phases:

$$Alm_2Cro_4(os) + 2OH^-(is) \leftrightarrow 2\,AlmOH(os) + Cro_4^{-2}(is) \tag{7}$$

$$K = \frac{[AlmOH]_{os}[Cro_4^{-2}]_{is}}{[Alm_2Cro_4]_{os}[OH^-]_{is}^2} \tag{8}$$

and following reaction:

$$H^+ + OH^- \leftrightarrow H_2O \qquad K = \frac{1}{k_w} = 10^{14} \tag{9}$$

(f) Diffusion of chromate ion to bulk of stripping phase:

$$Ncro_4^{-2} = \frac{Ds}{Dis}\left([Cro_4^{-2}]_{is} - [Cro_4^{-2}]_s\right) \tag{10}$$

2.8 DIPHENYLCARBAZIDE METHOD

Diphenylcarbazide (sym-diphenylcarbazide, diphenylcarbohydrazide) reacts in acid medium with Cu (II) ions to give a violet solution which is the basis of this sensitive method.

Many investigators have studied the reaction [8], offering rather divergent explanations of its mechanism. Pflaum and Howick [9], among others [12, 15, 16], have shown that the cationic (II) and diphenylcarbazone fails to yield a violet color. In all probability, the reaction involves unhydrated chromium (III) ions formed during the oxidation of diphenylcarbazide to diphenylcarbazone.

This explanation is however, incomplete since, when the colored reaction product is extracted into isoamyl alcohol or chloroform in the presence of perchlorate, the remaining colorless aqueous phase contains half of the chromium [10, 13]. When studying the reactions of diphenylcarbazide and diphenylcarbazone with various metal cations, Balt and Van Dalen [18] found that diphenylcarbazide only forms metal chelates after its oxidation to diphenylcarbazone.

2.9 RESULTS

2.9.1 Result of Feed Phase for Effect of pH

(Molarity of stripping phase = 3M, rotational speed of mixer in all phases = 100 rpm, and volumetric percentage of carrier = 1%, with presence of dodecanol in membrane phase).

Time (min)	Concentration(ppm) pH = 2	Concentration(ppm) pH = 4	Concentration(ppm) pH = 6
1440	0.1 ± 0.0	0.3 ± 0.1	0.5 ± 0.2

2.9.2 Result of Stripping Phase for Effect of pH

(Molarity of stripping phase = 3M, rotational speed of mixer in all phases = 100 rpm, and volumetric percentage of carrier = 1%, with presence of dodecanol in membrane phase).

Time (min)	Concentration(ppm) pH=2	Concentration (ppm) pH=4	Concentration (ppm) pH=6
1440	70.4 ± 0.2	6.4 ± 0.8	58.6 ± 0.6

2.9.3 Result of Feed Phase for Effect of Percentage of Carrier in Membrane Phase

(Molarity of stripping phase = 3M, rotational speed of mixer in all phases = 100 rpm, and pH = 2, with presence of dodecanol in membrane phase).

Time (min)	Concentration (ppm) % carrier $\left(\frac{v}{v}\right) = 0.5$	Concentration (ppm) % carrier $\left(\frac{v}{v}\right) = 1$	Concentration (ppm) % carrier $\left(\frac{v}{v}\right)=2$
1440	5.6 ± 0.5	0.1 ± 0.0	0

2.9.4 Result of Stripping Phase for Effect of Percentage of Carrier in Membrane Phase

(Molarity of stripping phase = 3M, rotational speed of mixer in all phases = 100 rpm, and pH = 2, with presence of dodecanol in membrane phase).

Time (min)	Concentration (ppm) % carrier $\left(\frac{v}{v}\right)$ = 0.5	Concentration (ppm) % carrier $\left(\frac{v}{v}\right)$ = 1	Concentration (ppm) % carrier $\left(\frac{v}{v}\right)$ = 2
1440	60.3 ± 0.4	70.4 ± 0.2	53.3 ± 0.5

2.9.5 Result of Feed Phase for Effect of Presence of Dodecanol in Membrane Phase

(Molarity of stripping phase = 3M, rotational speed of mixer in all phases = 100 rpm, and pH = 2, % Carrier $\left(\frac{v}{v}\right)$ = 1).

Time (min)	Concentration (ppm) with presence of dodecanol in membrane	Concentration without presence of dodecanol in membrane phase
1440	0.1 ± 0.0	6.4 ± 0.4

2.9.6 Result of Stripping Phase for Effect of Presence of Dodecanol in Membrane Phase

(Molarity of stripping phase = 3M, rotational speed of mixer in all phases =100 rpm, and pH = 2, % Carrier $\left(\frac{v}{v}\right)$ = 1).

Time (min)	Concentration (ppm) with presence of dodecanol in membrane	Concentration(ppm) without presence of dodecanol in membrane phase
1440	70.4 ± 0.2	52.6 ± 0.9

2.9.7 Result of Feed Phase for Effect of Velocity of Mixer in Stripping Phase

(Molarity of stripping phase = 3M, rotational speed of mixer in feed phase= 100 rpm, and volumetric percentage of carrier = 1%, with presence of dodecanol in membrane phase, pH = 2).

Time(min)	Concentration(ppm) velocity of mixer in stripping phase = 80 rpm	Concentration(ppm) velocity of mixer in stripping phase = 100 rpm	Concentration(ppm) velocity of mixer in stripping phase = 120 rpm
1440	5.7 ± 0.5	01 ± 0.0	3.7 ± 0.4

2.9.8 Result of Stripping Phase for Effect of Velocity of Mixer in Stripping Phase

(Molarity of stripping phase = 3M, rotational speed of mixer in feed phase = 100 rpm, and volumetric percentage of carrier = 1%, with presence of dodecanol in membrane phase, pH = 2).

Time(min)	Concentration(ppm) velocity of mixer in stripping phase = 80 rpm	Concentration(ppm) velocity of mixer in stripping phase = 100 rpm	Concentration(ppm) velocity of mixer in stripping phase = 120 rpm
1440	56.4 ± 0.9	70.4 ± 0.2	60.6 ± 0.9

2.9.9   Result of Feed Phase for Effect of Velocity of Mixer in Feed Phase

(Molarity of stripping phase = 3M, rotational speed of mixer in stripping phase = 100 rpm, pH = 2, and volumetric percentage of carrier = 1%, with presence of dodecanol in membrane phase).

Time(min)	Concentration(ppm) velocity of mixer in stripping phase =80 rpm	Concentration(ppm) velocity of mixer in stripping phase =100 rpm	Concentration(ppm) velocity of mixer in stripping phase =120 rpm
1440	6.2 ± 0.4	0.1 ± 0.0	3.0 ± 0.4

2.9.10   Result of Stripping Phase for Effect of Velocity of Mixer in Feed Phase

(Molarity of stripping phase = 3M, rotational speed of mixer in stripping phase = 100 rpm, pH = 2, and volumetric percentage of carrier = 1%, with presence of dodecanol in membrane phase).

Time(min)	Concentration(ppm) velocity of mixer in feed phase =80 rpm	Concentration.3ppm) velocity of mixer in feed phase =100 rpm	Concentration(ppm) velocity of mixer in feed phase =120 rpm
1440	54.7 ± 0.5	70.4 ± 0.2	57.1 ± 0.7

2.9.11   Result of Feed Phase for Effect of Molarity of Stripping Phase

(Rotational speed of mixer in all phases = 100 rpm, pH = 2, and volumetric percentage of carrier = 1%, with presence of dodecanol in membrane phase).

Time(min)	Concentration(ppm) 1 M Solution	Concentration(ppm) 2 M Solution	Concentration(ppm) 3 M Solution
1440	0.2 ± 0.7	6.2 ± 0.4	6.2 ± 0.4

2.9.12   Result of Stripping Phase for Effect of Molarity of Stripping Phase

(Rotational speed of mixer in all phases = 100 rpm, pH = 2, and volumetric percentage of carrier = 1%, with presence of dodecanol in membrane phase).

Time(min)	Concentration(ppm) 1 M Solution	Concentration(ppm) 2 M Solution	Concentration(ppm) 3 M Solution
1440	56.5 ± 0.7	66.5 ± 0.6	70.4 ± 0.2

2.9.13   Chromium Accumulation % in Membrane Phase

$$(C.A) = \frac{A - (B - C)}{A}$$ (see Table 1)

where
A = Initial moles of chromium in feed phase
B = Final moles of chromium in feed phase
C = Final moles of chromium in stripping phase

TABLE 1   Percentage of chromium in membrane phase.

Parameter	Amount	C. A
	2	10.66
pH of feed phase	4	14.44
	6	22.44
	0.5	4.44
% Carrier(v/v)	1.0	10.66
	2.0	33.33
Presence/absence of dodecanol in membrane phase	YES	10.66
	NO	8.88
	1	26.66
Molarity of stripping phase	2	14.00
	3	10.66
	80	4.44
Velocity of mixer in feed phase	100	10.66
	120	15.55
	80	15.55
Velocity of mixer in stripping phase	100	10.66
	120	20.88

2.9.14   Chromium Extraction Percentage for Each Parameter

Extraction % $= C/A * 100$ (see Table 2)
where,
A = Initial moles of chromium in feed phase
C = Final moles of chromium in stripping phase

TABLE 2   Extraction percentage of chromium for each parameter.

Parameter	Amount	Extraction %
	2	88.88
pH of feed phase	4	74.44
	6	71.11
	80	71.11
Velocity of mixer in feed phase	100	88.88
	120	73.33
	80	64.44
Velocity of mixer in stripping phase	100	88.88
	120	77.77
	1	71.11
Molarity of stripping phase	2	84.44
	3	88.88
Presence/absence of dodecanol in membrane phase	YES	88.88
	NO	66.66
	0.5	75.55
% Carrier(v/v)	1.0	88.88
	2.0	66.66

2.10   DISCUSSION

2.10.1   Effect of Presence/Absence of Surfactant in Membrane Phase

Presence of surfactant in membrane phase decreases surface tension between liquor and organic phases and because of this reason, mass transfer increases. Surfactant should be ineffective, has a high solubility in organic phase and has a polarity in one end and 5 ml dodecanol was used in experiments. Using high amounts of this material create jelling state in surface of liquor and organic phases. Also when high amounts of dodecanol were used, Alamine salts: surrounded by dodecanol and cannot react immediately, therefore extraction decreases. As we see in Figure 2, After 24 hr in our experiment with presence of dodecanol, concentration of chromium in feed phase is 0.1 ppm and concentration of chromium in the experiment without presence of dodecanol after 24 hr is 6.4 ppm. Also in Figure 3, Concentration of chromium in stripping phase with presence of dodecanol is 70.4 ppm.

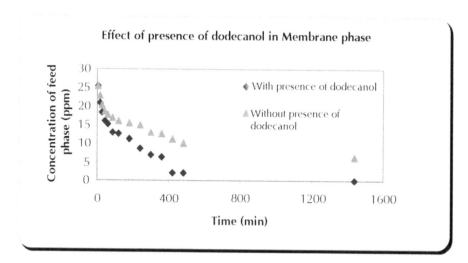

FIGURE 2   Concentration of chromium in feed phase with presence of dodecanol.

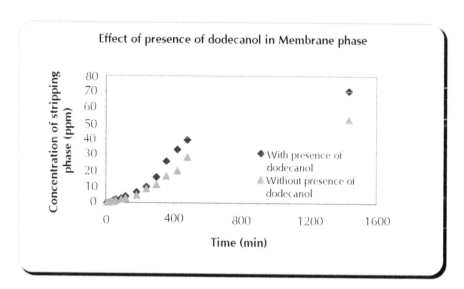

FIGURE 3   Concentration of chromium in stripping phase with presence of dodecanol.

2.10.2   Effect of pH

Decreasing of pH in feed phase cause to neutralize OH in Formula (2) and reaction goes to right side: it increases speed of this formula. Therefore, if pH was bounded on two, membrane efficiency increases. If pH was decreased so much may be membrane and carrier was oxidized, because of oxidizing property of chromate (Figure 4).

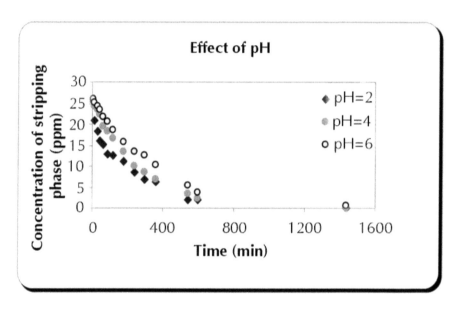

FIGURE 4   Decreasing of pH in feed phase.

2.10.3   Effect of Volumetric Percentage of Carrier

With increasing of percentage carrier in membrane phase, extraction increases and in definite concentration of carrier (1%) maximum percentage of extraction was obtained. But after that, because of increasing in amount, in membrane phase and also increasing of carrier, extraction decreases. Then optimum amount of carrier should be defined. With increasing of percentage carrier in membrane phase, extraction was operated successfully but because of increasing of chromium in membrane phase, back extraction (extraction from organic phase to stripping phase) was operated slowly. As we considered percentage carrier was played important role in extraction process. Reaction (2) is faster than Reaction (7); therefore, amount of entering chromium to membrane phase is more than amount of balcony chromium from membrane phase. Since chromium decreases in feed phase (Figure 5) in one an exact time speed of entering and balconing chromium equal to each other and amount of chromium in membrane phase goes maximum (Figure 6).

Then optimum carrier percentage is so important in extraction processes, and in our experiments it was 1% (V/V).

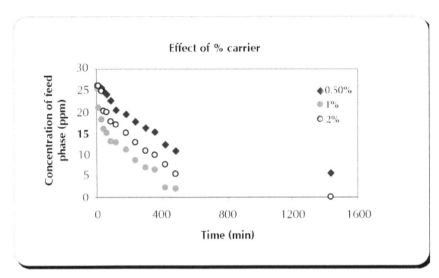

FIGURE 5   Effect of percentage carrier in concentration of feed phase.

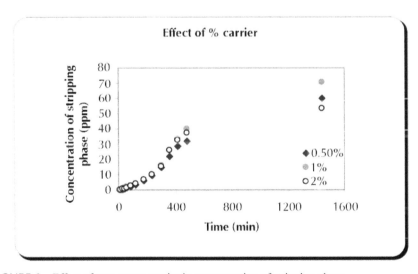

FIGURE 6   Effect of percentage carrier in concentration of stripping phase.

2.10.4   Effect of Mixing

The speed of mixing of phases is so important that the fastest speed of mixing of phases lead to smallest thickness of films in Equations (1), (5), (6), and (10) because of this reason, mass transfer decreases. As we see in diagrams, the optimum speed of mixing is 100 rpm. Because in some other speed like 80 rpm, flow is laminar and it is so silent and phases cannot mix together as they can, and mass transfer do not operate successfully.

Because of these reason extraction decreases. The next speed of mixing of phases that we investigated was 120 rpm. In 120 rpm, flow is turbulent because of this reason, fine bubbles of feed and stripping phases enter to membrane phase and because of high velocity of mixers these bubbles cannot operate inter facial reactions and go straight to the other liquor phase and extraction decreases. As we see in Figure 7, concentration of feed phase after 24 hr, in 80, 100, and 120 rpm approaches to 0.1, 3, and 6.2 ppm respectively. Also in Figure 8, extraction in 100 rpm, is higher than the others. And after 24 hr concentration of stripping phase in 80, 100, and 120 rpm is 56.4, 60.6, and 70.4 ppm respectively.

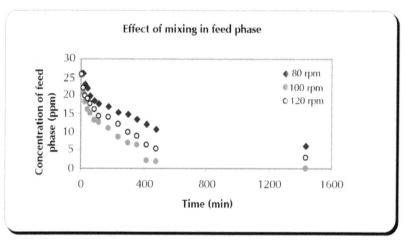

FIGURE 7   Effect of mixing concentration in feed phase.

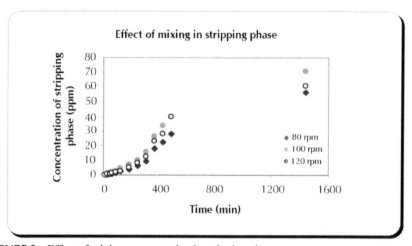

FIGURE 8   Effect of mixing concentration in stripping phase.

2.10.5 Effect of Molarity of Stripping Phase

Because of low concentration of chromium in feed phase and pay attention to this point that with each chromate ion, two H^+ ions were transferred from feed phase to stripping phase variation of pH in stripping phase is negligible. Therefore, most basic stripping phase lead to operate extraction easily. And we see this point in the (Figure 9) and results.

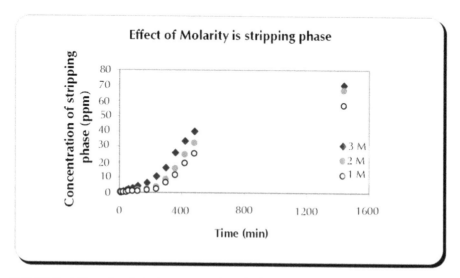

FIGURE 9 Effect of molarity in concentration of stripping phase.

2.11 CONCLUSION

(1) Presence of surfactant in membrane phase increased extraction process.

(2) Optimum amount of carrier percentage in our experiments was 1% (V/V). In this measure, maximum extraction was obtained.

(3) Optimum amount of pH for feed in our experiments was pH = 2. In this pH, extractions are increased and Alm_2CrO_4 complexes are generated faster than the other conditions that we did.

(4) Optimum velocity of mixer for both liquor phases (feed phase and stripping phase) in our experiments is 100 rpm. Because extraction and back extraction are operated successfully.

(5) Best molarity for stripping phase in our experiments is Molarity = 3 because under this condition back extraction are operated successfully and AlmoH Complexes are generated faster than the other conditions that we did.

KEYWORDS

- **Diphenylcarbazide method**
- **Dodecanol**
- **Feed phase**
- **Membrane phase**
- **Stripping phase**

REFERENCES

1. Douglas Way, J., Richard D. Noble, Thomas M. Flynn, and Dendy Sloan, E. *Journal of Membrane science*, **12**, 239–259 (1982).
2. Schultz, J. S., Goddard, J. D., and Suchdeo, S. R. Facilitated transport via carrier-mediated diffusion in membranes Part I. *AIChE J.*, **20**, 417 (1974a).
3. Schultz, J. S., Goddard, J. D., and Suchdeo, S. R. Facilitated transport via carrier-mediated diffusion in membranes Part II. *AIChE J.*, **20**, 625 (1974b).
4. Smith, D. R., Lander, R. J., and Quinn, J. A. Carrier-Mediated Transport in Synthetic Membranes. In *Recent Developments in Separation Science.*, *Vol 3.* N. N. Li (Ed.), CRC Press, Cleveland, Ohio (1977).
5. Kimura, S. G., Matson, S. L., and Ward, W. J. Industrial Applications of Facilitated Transport. In *Recent Developments in Separation Science, Vol 5.* N. N. Li (Ed.), CRC Press, Cleveland, Ohio (1979).
6. Goddard, J. D. Further applications of carrier mediated transport theory–A survey. *Chem. Eng. Sci.*, **32**, 795 (1977).
7. Halwachs, W. and Schugerl, R. The liquid membrane technique–A promising extraction process. *Int. Chem. Eng.*, **20**, 519 (1980).
8. Bose, M. ibid., **10**, 201–209 (1954).
9. Pflaum, R. T. and Howick, L. C. *Am. J. Chem. Soc.*, **78**, 4862 (1956).
10. Lichtenstein, I. E. and Allen, T. L. ibid: **81**, 1040, *J. Phys. Chem.*, **65**, 1238 (1961).
11. Babko, A. K., Get'man, T. E. *Zh. Obshch. khim.*, **29**, 2416 (1959).
12. Minczewski, J. and Zemijewska, W. *Roczniki Chem.*, **34**, 1559, *Chem. Anal.*, **5**, 429 Warsaw (1960).
13. Sano, H. *Anal.Chim.Acta*, **27**, 398 (1962).
14. Kemula, W., Kublik, Z., and Najdeker, E. *Roczniki Chem.*, **36**, 937 (1962).
15. Kovalenko, E. V. and Petrashen, V. I. *Zh. Analit. Khim*, **18**, 743 (1963).
16. Zittel, H. E. *Anal.Chem.*, **35**, 329 (1963).
17. Allen, T. L. *Anal.Chem.*, **30**, 447 (1958).
18. Balt, S. and Van Dalen, E. *Anal. Chim. Acta*, **25**, 507 (1961).

3 Excess Permittivity for Mixtures at Various Concentrations: An Experimental Approach

CONTENTS

3.1 INTRODUCTION

New experimental data are reported for the thermodynamic investigation of the inter-molecular halogenated compounds in non-polar solvent. Experimental data are reported of excess molar volumes, excess permittivity, and excess Gibb's free energy (ΔG) of binary mixtures of Flourobenzene, Chlorobenzne, Bromobenzne, 4-Chlorobenzal-dehyd, and 4-flourobenzaldehyd in 1,4-dioxane at various concentrations at 298, 308, and 318K. The mixing calculated using the experimental value of relative permittivity on the basis of Winkelmann-Quitzsch equation. The V^E values were measured using a dilatometer and are positive over the entire mole fraction range for all the systems.

Experimental data of excess thermodynamic properties of liquids and liquid mixtures are fascinating and of great fundamental, practical importance, and industrial points of view. Moreover, these properties allow one to draw information on the structure and interactions of mixed solvents. The chemical industries have recognized the importance of the thermodynamic properties in design calculations involving chemical separations, heat transfer, mass transfer and fluid flow. The static dielectric measurement has been shown to be useful technique in characterizing molecular interaction and molecular ordering in solutions [1-3]. Experimental data of excess thermodynamic properties of liquids and liquid mixtures are fascinating and of great fundamental, practical importance, and industrial points of view. Moreover, these properties allow one to draw information on the structure and interactions of mixed solvents. We have also measured the relative permittivity, ε, the refractive indices, n_D, and the densities, ρ, of halogenated compounds and liquid crystal.

3.2 EXPERIMENTAL

3.2.1 Material

Flour benzene, Chlorobenzene, Bromobenzene, 4-chlorobenzaldehyd, and 4-flourobenzaldehyd were obtained from Merck at a purity of 99.8%. All compounds studied in this investigation are listed in Table 1.

3.2.2 Dielectric Apparatus

The electrical capacitance of the dielectric cell was measured using a Wayne Kerr model 6425B Digibridge. Measurements of the capacitance was performed at a frequency of 10 kHz.

3.2.3 Dielectric Cell

A cylindrical dielectric cell was constructed for measurements on small volume of solution (2.5 ml). Measurements of the capacitance required for calculating the static dielectric permittivity were performed at a frequency of 10 kHz. The electrical capacitance of the empty cell was about 30 pF. A water circulated thermostat was used to maintain the temperature at 25, 35, and 45°C.

TABLE 1   Compounds use in this study.

Compound	System	Abbreviation
⬡—F	Flour benzene	FB
⬡—Cl	Chlorobenzene	ClB
⬡—Br	Bromobenzene	BrB
OHC—⬡—F	4-Flourobenzaldehyd	4FBA
OHC—⬡—Cl	4-Chlourobenzaldehyd	4ClBA

3.2.4   Refractive Index Measurement

The refractive indices of the solutions were at a wavelength of 589 nm using an Abbe refractometer (Model CETI). The temperature of the refractometer was controlled by circulation of water.

3.2.5   Density Measurement

The densities of the solutions were measured using a 5 ml specific gravity bottle. The error in the measurement of density was $\pm$ 0.0002 mg/cc.

3.3   DISCUSSION AND RESULT

The excess molar volumes of the solutions of molar composition x were calculated from the densities of the pure liquids and their mixtures according to the following Equation [4]

$$V^E = [xM_1 + (1-x)M_2]/\rho - [xM_1/\rho_1 + (1-x)M_2/\rho_2]$$

where, ρ, ρ_1, and ρ_2 are the densities of the solution and pure components 1 and 2 respectively, and M_1 and M_2 are the molar masses of the pure components. The corresponding V^E values of halogenated compounds at 298.15, 308.15, and 318.15K are shown in Figure 1, Figure 2, and Figure 3.

The dipole moment of Bromobenzen is 1.45 D ,which is smaller than Chloroben-zen and 4-florobenzaldehyd, so it is considered that the dipole-dipole interactions of Chlorobenzen and 4-florobenzaldehyd with solvent is weaker than those of Bromobenzen with 1,4-dioxane. This difference in dipole moment between Chlorobenzen and 4-florobenzaldehyd and Bromobenzen may explain, to some extent, the fact that the V^E (x = 0.5) value for (Chlorobenzen and 4-florobenzaldehyd + 1,4-dioxane) is larger than that for (Bromobenzen + 1,4-dioxane) in mixtures. Generally, a change in volume upon mixing can be attributed to various factors [7-9], which are usually (a) differ-ences in sizes and shapes of the molecules of the pure components, (b) differences in the intermolecular interaction energy between like and unlike species, and (c) the formation in the mixture of a chemical complex composed of unlike molecules.

The information related to the heterogeneous interaction may also be obtained by the value of excess permittivity [5].

$$\varepsilon^E = (\varepsilon_0 - \varepsilon_\infty)_m - [(\varepsilon_0 - \varepsilon_\infty)_1 X_1 + (\varepsilon_0 - \varepsilon_\infty)_2 X_2]$$

The excess permittivity provides qualitative information about multimer formation in the mixture.

The excess permittivity ε^E also provides information about the formation of mu-timers in the mixtures. The variation of ε^E with the mol fraction X_2 of the halogenated compounds at 25, 35, and 45°C is shown by the plots for in Figure 4.

Deviation in ε^E from zero is a measure of the strength of the intermolecular interac-tion between solute and solvent in the mixture.

If $\varepsilon^E > 0$ then the interaction between solute and solvent is such that the effective dipole moment in the mixture increases with parallel orientation among the dipoles. If $\varepsilon^E < 0$, then the solute-solvent interaction is such as that the effective dipole moment in the mixture gets reduced giving rise to antiparallel orientation among the dipoles.

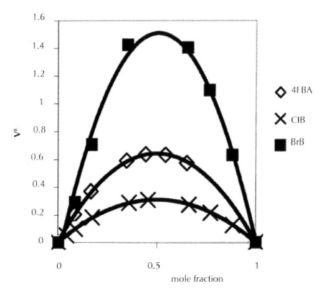

FIGURE 1   Excess molar volume Bromobenzene, Chlorobenzen, and 4-florobenzaldehyd at 298K.

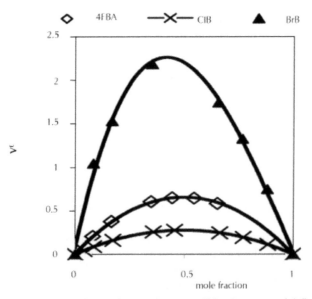

FIGURE 2   Excess molar volume of Bromobenzene, Chlorobenzen, and 4-florobenzaldehyd at 308K.

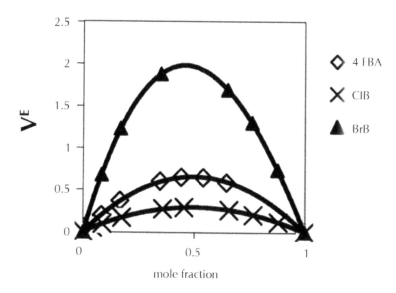

FIGURE 3   Excess molar volume Bromobenzene, Chlorobenzene, and 4-florobenzaldehyd at 318K.

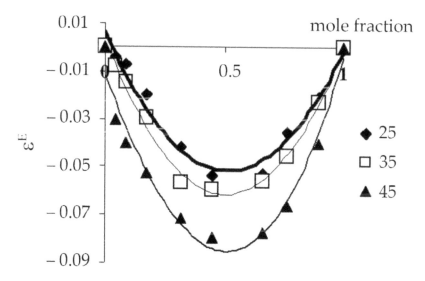

FIGURE 4   Excess permittivity (ε^E) versus mole fraction (X_2) of Chlorobenzene in 1,4-dioxan.

The expression for excess Gibb's energy (ΔG) for a matter in condensed phase is given by [6]

$$\Delta G = -\frac{N}{2}[R_{f2} - R^0{}_{f2}][X_2 \mu^2 (X_2 (g'-1)+1)]$$

where

$$R^0{}_{f2} = \frac{8\pi N(\varepsilon_2 - 1)(\varepsilon_\infty + 2)}{9V_2(2\varepsilon_2 + \varepsilon_\infty)} \qquad R_{f2} = \frac{8\pi N(\varepsilon_m - 1)(\varepsilon_\infty + 2)}{9V_2(2\varepsilon_m + \varepsilon_\infty)}$$

$\varepsilon_{0m,}$ and ε_{02} respectively are the values of static permittivity corresponding to mixture and pure solute. The $R^0{}_{f2}$, R_{f2} are reaction field parameter in the pure liquid as that in the mixtures and g' is linear correlation factor of pure solute.

The trend of variation of excess Gibb's free energy ΔG in mixtures seems to be identical in all mole fraction remain positive throughout.

It is found that excess free energy increases with the increase in mole fraction of solute which attains maximum at approximately 0.5 mole fraction of solute and finally drops to zero.

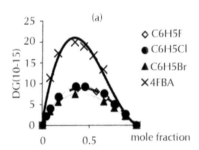
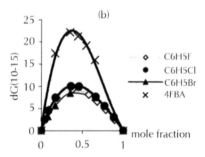

FIGURE 5   Variation of DG with mole fraction Florobenzene, Chlorobenzene, Bromobenzene, and 4-florobenzaldehyde at (a) 25°C and (b) 35°C.

3.4   CONCLUSION

For three temperatures, values of permittivity, excess permittivity for mixtures at various concentrations are reported. The nature of the molecular interaction and molecular association between the components in the mixture at concentration at different temperatures is identified and results are interpreted. The temperature derivatives of the excess volumes are also positive and large.

KEYWORDS

- **Chlorobenzene**
- **Dielectric cell**
- **Excess permittivity**
- **Gibb's energy**
- **Winkelmann-Quitzsch equation**

REFERENCES

1. Ghanadzadeh, A., Mamaghani, M., and Anbir, L. *J. Sol. Chem.*, **32**(7), 365 (2003).
2. Ghanadzadeh, A. and Beevers, M. S. *J. Mol. Liq.*, **100**(1), 47, 365 (2002, 2003).
3. Ghanadzadeh, A. and Beevers, M. S. *J. Mol .Liq.*, **102**, 365 (2003).
4. Redlich, O. J. and Kister, A. T. *Ind. Eng. Chem.*, **40**, 345–348 (1948).
5. Tabellout, M., Lancoleur, P., Emery, J. R., Hayward, D., and Pethrick, R. A. *J. Chem. Soc., Faraday Trans.*, **86**, 1453 (1990).
6. Winkelmann, J. and Quitztch, K. *Z. Phys. Chem.*, **257**, 746 (Leipzig) (1976).
7. Pal, A. and Singh, W. *J. Chem. Thermodyn.*, **28**, 227–232 (1996).
8. Pal, A. and Singh, W. *Fluid Phase Equilibria.*, **129**, 211–221 (1997).
9. Belousov, V. P. and Panov, M. Y. In *Thermodynamic Properties of Aqueous Solution of Organic Substances*. CRC Press, London (1994).

4   Predict Natural Gas Water Content

CONTENTS

4.1 INTRODUCTION

Natural gas reservoirs always have water associated with them: gas in the reservoir is saturated by water. When the gas is produced water is produced too from the reservoir directly. Other water produced with the gas is water of condensation formed because of the changes in pressure and temperature during production. In the transmission of natural gas further condensation of water is troublesome [1]. It can enlarge pressure drop in the line and frequently goes to corrosion problems. Therefore, water should be removed from the natural gas before it is offered to transmit in the pipeline. For these argue, the water content of sour gas could be important for engineering attention. In a study of the water content of natural gases Lukacs [1] measured the water content of pure methane at 160°F and pressures up to 1,500 psia also Gillespie et al. [2] predicted the water content of methane in the range of 122–167°F and for pressures from 200 to 2,000 psia. Sharma et al. [3] proposed a method for calculating the water content of sour gases, originally designed for hand calculations, but it was slightly complicated. Bukacek [4] suggested a relatively simple correlation for the water content of sweet gas, based on using an ideal contribution and a deviation factor. McKetta et al. published a chart for estimating the water content of sweet natural gas. This chart has been modified slightly over the years and has been reproduced in many publications [5]. Recently, Ning et al. [6] proposed a correlation based on the McKetta et al. chart. This correlation reveals how difficult it can be to correlate something that is as seemingly simple as the water content of natural gas. Maddox [7] developed a method for estimating the water content of sour natural gas. His method assumes that the water content of sour gas is the sum of three terms sweet gas contribution (Methane, CO_2, and H_2S).

Most of the traditional methods work in the limited range of pressure and temperature and they have a good accuracy in this limited range, which is near the ideal equilibrium condition. But in the high pressure and temperature gases have nonlinear behavior that these methods cannot predict the gas behavior [10]. The ANN as a good nonlinear function approximator can simulate the nonlinear functions with high accuracy [14, 16]. In this chapter we predicted the water content of the sour natural gas mixtures with the ANN method. The results show the ANN's capability to predict the measured data. We compare our results with the other numerical and analytical methods. For example Wichert and Bukacek Maddox. These comparisons confirm the superiority of the ANN method.

In this chapter a new method based an (ANN) for prediction of natural gas mixture water content is presented. The dehydration of natural gas is very important in the gas processing industry, for design of facilities of the production, transmission, and processing of natural gas. It is necessary to remove water vapor present in the gas stream that may cause hydrate formation at low temperature conditions that may plug the valves and fittings in gas pipelines. In this study, the available data for mixtures and available methods for predicting the water content of sour gas have been studied. Based on obtained results from ANN simulation, our methods is more accurate than current used methods and can be used in gas engineering studies.

4.2   METHOD

4.2.1   Artificial Neural Networks

The ANN is constructed as a massive connection model of simply designed computing unit called "neuron". Figure 1 illustrates a simple model of n-inputs single-output neuron. All the input signals are summed up as z and the amplitude of the output signal is determined by the nonlinear activation function $f(z)$. In this work, we employ the modified sigmoid function $f(z)$ given as follow [14]:

$$f(z) = \frac{1 - e^{-kz}}{1 + e^{-kz}} \tag{1}$$

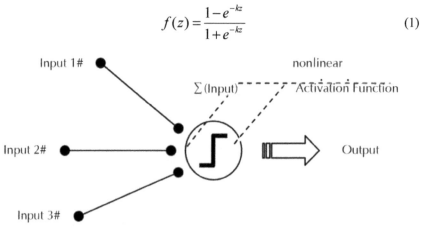

FIGURE 1   Basic model of multi-inputs one-output neuron.

Here, we adopt the sigmoid function with moderate slope so that the network can output continuous range of values from −1 to 1, which brings the differentiability of the network [14, 15]. Here, we adopt a multilayer perceptron (MLP) type network with three layers, which has been used for various applications [14-16]. Shown in Figure 2 is the architecture of the perceptron neural network. For clear notation, we will use the indices i, j and k for the units corresponding to "input", "hidden", and "output" layers, respectively (see Figure 2). Note also that n_i and o_i are used to represent the input and output to the ith neuron, respectively. Input-output properties of the neurons in each layer can be simply expressed in mathematical term as [16]:

$$o_i = f(n_i), \tag{2}$$

$$o_j = f(n_j), \tag{3}$$

$$o_k = f(n_k), \tag{4}$$

whereas, inputs to the neurons are given as:

$$n_i = (\text{input signal to the ANN}), \tag{5}$$

$$n_j = \sum_{i=1}^{N_i} w_{ij} o_i + \theta_j, \tag{6}$$

$$n_k = \sum_{j=1}^{N_j} w_{jk} o_j + \theta_k. \tag{7}$$

Here, N_i and N_j represent the numbers of the units belonging to "input" and "hidden" layers, while w_{ij} denotes the synaptic weight parameter which connects the neurons i and j.

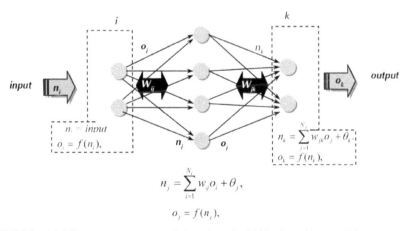

FIGURE 2   Multilayer perceptron consisting "input", "hidden", and "output" layers.

The ANN training is an optimization process in which an error function is minimized by adjusting the ANN parameters (weights and biases). When an input training pattern is introduced to the ANN, it calculates an output. Output is compared with the real output (experimental data) provided by the user.

We train the network *via* the fast convergence gradient descend back-propagation method with momentum term for the nonnegative energy function [12].

4.3   RESULTS

In this part of our study, the object is to find the optimal performance ANN model for prediction of water content. The results are illustrated in Figure 3. In this work, an according Figure 3, the optimum number of hidden nodes was selected to be 302. In selecting data for modeling, and to ensure that they represent normal operating ranges, off data were deleted from the data list. The variables of the model and the operating ranges are summarized in Table 1.

The data sets were collected from various components in this simulation. For an ANN simulation of gas mixture water content, data sets obtained by Ng et al. [6], Lukacs [1], GPSA Engineering Data Book [5]. These data sets are an important sets and suitable for achieving studies on the behavior of natural gases.

Major components used in these references are methane, propane, hydrogen sulfide, carbon dioxide, and water. The analysis data are based on sour gas. Overall 136 data sets were obtained. Among 136 data sets 80 points were used for training the ANN and the remaining 56 data sets were used for accuracy checks of the simulation. The input variables of model and their operating ranges are limited to hydrogen sulfide composition up to 89.6 mol% and are applicable for temperatures between 50 and 350°F and pressure from 200–3,500 psia. Finally, the water content as output of ANN at the range of 40.6 to 3,500 lb/MMCF.

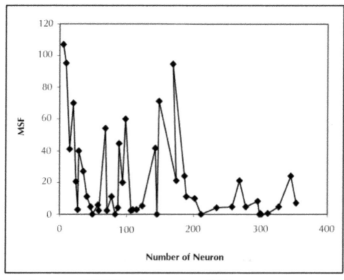

FIGURE 3   Performance of ANN based on number of hidden neurons.

Wichert correlation [8, 9] is valid only for H_2S up to 55 mol% in sour gas mixtures. This only covers 68% of data of valid experimental data. In this simulation of gas mixture water content we compared the results of ANN by Wichert and Bukacek-Maddox (BM). At this compare BM and Wichert could not predict sour natural gas water content in a wide range of data. These methods have limitations and only can be used in an appointed pressure and temperature. But ANN can predict unknown behaviors of sour natural gases in any limitations of pressure and temperature conditions.

TABLE 1 ANN model variables and their ranges.

Variable	Range
H_2S (%mol)	7.96–89.52
T (F)	50–350
P (psia)	200–3,500
Water Content (lb/MMCF)	40.6–3,500

TABLE 2 Comparison of the ANN prediction results of the sour natural gas mixture water content with the Wichert and Bukacek-Maddox methods.

H_2S % mol	T (F)	P (psia)	Experimental	
			W.C (calc)	Div (%)
7.96	200.0	200	2835.0	2506.0
8.00	130.0	1500	111.0	116.0
9.06	200.0	200	2820.0	2500.0
10.00	100.0	1100	81.0	75.0
15.71	120.0	200	414.8	380.1
16.00	159.8	1395	226.0	231.0
17.00	160.0	1010	292.0	294.0
17.46	120.0	200	526.5	379.2
18.10	120.0	200	378.8	378.6
19.00	160.0	611	442.0	418.0
21.00	160.0	358	712.0	707.0
27.50	160.0	1392	247.0	264.0

TABLE 2   *(Continued)*

H$_2$S % mol	T (F)	P (psia)	Experimental	
			W.C (calc)	Div (%)
27.50	160.0	1367	247.0	268.0
29.00	160.0	925	328.0	330.0
43.80	120	200	568.4	388.0
47.30	200	200	3087.0	2462.0
75.56	120	200	559.1	....
81.25	200	200	2916.0	....
89.52	120	200	619.9	....

AAD %

Wichert	BM		ANN	
	W.C (calc)	Div (%)	W.C (calc)	Div (%)
11.60	2866.0	-1.09	2890.0	-1.94
-4.50	113.0	-1.80	113.0	-1.80
11.34	2855.0	-1.24	2877.9	-2.05
7.41	83.0	-2.47	81.3	-0.37
8.36	430.5	-3.66	420.3	-1.33
-2.21	260.0	-15.04	231.9	-2.61
-0.68	322.0	-10.27	298.6	-2.26
32.59	433.7	17.62	529.4	-0.55
0.05	434.2	-14.62	384.9	-1.66
5.43	467.0	-5.66	447.0	-1.13
0.07	723.0	-1.54	709.3	0.38
-6.88	297.0	-20.24	253.3	-2.55

TABLE 2   *(Continued)*

AAD %

Wichert	BM		ANN	
	W.C (calc)	Div (%)	W.C (calc)	Div (%)
-0. 61	375.0	-0. 61	334.6	-2.01
31.74	446.0	21.53	570.1	-0.30
20.25	2900.0	6.06	3070.1	0.55
....	461.7	17.42	561.1	-0.36
....	2941.0	-0.86	2954.0	-1.30
....	463.3	25.26	620.0	-0.02
9.514	9.917		1.437	

4.4   DISCUSSION

A scatter plot of measured water content against the ANN model predictions is show
in Figure 4. The prediction, which match measured values, should fall on the diagonal
line (line with intercept 0 and slope equal to 1). Almost all data lay on this line, which
can confirms the accuracy of the ANN model.

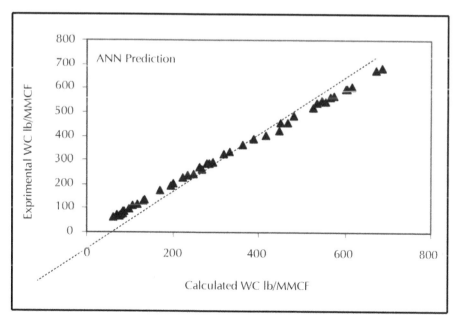

FIGURE 4   Artificial neural network prediction of gas mixtures water content (lb/MMCF).

The data points are very close to the diagonal lines and this confirms again the ANN can learn very well relationships between input and output data and generalized successfully of water content. Good performance of ANN is obvious when it is compared to other models and simulators. To check the performance of the ANN model, its estimation and compared with an existing simulator available. The Table 2 compares the error of ANN model with Wichert and BM models.

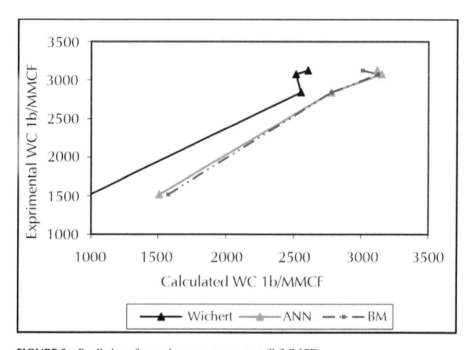

FIGURE 5 Prediction of gas mixture water content (lb/MMCF).

Figure 5 compares methods at the range of 1,000–3,500 lb/MMCF of water content. Error of Wichert calculations at this area is more than BM and ANN calculations.

4.5 CONCLUSION

The ANN model is developed for prediction of natural sour gas mixtures water content. The model is trained based on measured (experimental) data for three various inputs, (H_2S, Pressure, and Temperature).

The difference between ANN model prediction and validation data was very small which confirmed the ability of ANN to accurately predict unseen data. The ANN model was also compared with two numerical and analytical models, Wichert and BM. The results showed that the ANN model accuracy outperform the traditional simulators.

KEYWORDS

- **Analytical method**
- **Artificial neural network**
- **Multilayer perceptron**
- **Natural gas**
- **Wichert and Bukacek-Maddox method**

REFERENCES

1. Lukacs, J. Water Content of Hydrocarbon – Hydrogen Sulphide Gases, MSc Thesis, Department of Chemical Engineering, University of Alberta, Edmonton, AB, (1962) and Lukacs, J. and Robinson, D. B. Water Content of Sour Hydrocarbon Systems. *Soc. Petrol. Eng. J.*, **3**, 293–297 (1963).

2. Gillespie, P. C., Owens, J. L., and Wilson, G. M. *Sour Water Equilibria Extended to High Temperature and with Inerts Present.* AIChE Winter National Meeting, Paper 34-b, Atlanta, GA, pp. 11–14, (March, 1984) and Gillespie, P. C. and Wilson, G. M. *Vapor-Liquid Equilibrium Data on Water-substitute Gas Components: N_2-H_2O, H_2-H_2O, CO-H_2O, H_2-CO-H_2O, and H_2S-H_2O.* Research Report RR-**41**, GPA, Tulsa, OK (1980).

3. Sharma, S. and Campbell, J. M. Predict Natural gas Water Content with Total Gas Usage. *Oil and Gas J.*, 136–137 (August 4, 1969).

4. Bukacek—quoted in McCain, W. D. *The Properties of Petroleum Fluids*, 2nd ed., PennWell Books, Tulsa, OK (1990).

5. *GPSA Engineering Data Book*, 11th ed., Gas Processors Suppliers Association, Tulsa, OK, (1998).

6. Ning, Y., Zhang, H., and Zhou, G. Mathematical Simulation and Program for Water Content Chart of Natural Gas (in Chinese). *Chem. Eng. Oil Gas*, **29**, 75–77 (2000).

7. Maddox, R. N. *Gas and Liquid Sweetening, 2nd ed.* M. John (Ed.). Campbell Ltd., pp. 39–42, (1974) and Maddox, R. N., Lilly, L. L., Moshfeghian, M., and Elizondo, E. *Estimating Water Content of Sour Natural Gas Mixtures.* Laurance Reid Gas Conditioning Conference, Norman, OK, (March, 1988).

8. Wichert, G. C. and Wichert, E. Chart Estimates Water Content of Sour Natural Gas. *Oil and Gas J.*, 61–64, (1993).

9. Robinson, J. N., Moore, R. A., and Wichart, E. Chart Help Estimate H_2O Content of Sour Gases. *Oil and Gas J.*, 76–78, (1978).

10. Hugan, S. S., Leu, A. D., and Robinson, D. B. The Phase Behavior of Two Mixture of Methane, Carbon, Dioxide, Hydrogen Sulfide, and Water. *Fluid Phase Equil.*, **19**, 21–23 (1985).

11. Saarineen, S., Bramley, R., and Cybenko, G. *The numerical solution of the neural network training problems.* CRSD report 1089, center for supercomputing research and development, university of Illinois, Urbana, (1991).

12. Yam, J. and Chow, T. A weight initialization method for improving training speed in feed-forward neural network, *Neuro computing*, **30**, 219–232 (2000).

13. Luo, Z. On the convergence of the LMS algorithm with adaptive learning rate for linear feedforward neural networks, *Neural Computation*, **3**(2), 226–245 (1991).

14. Hagan, M. T. and Menhaj, M. B. Training Feed-forward Neural Network with the Marquardt Algorithm. *IEEE Transaction on Neural Networks*, **5**(6) (November, 1994).

15. Chong, E. K. P., Hui, S., and Stanislaw H. Zak,. An analysis of a class of neural networks for solving linear programming problems. *IEEE transactions on automatic control*, **44**(11) (1999).

16. Haykin, S. *Neural Networks: A Comprehensive Foundation, 2nd ed.* Prentice-Hall, New York (1999).

5 Optimization of Process Variables

CONTENTS

5.1 INTRODUCTION

Butanediol is a colorless and odorless liquid chemical with a very high boiling point and low freezing point. It is largely used as a monomer for polymer synthesis. The commercial applications of this diol are not limited to the manufacture of butadiene, or to its use as an antifreeze agent [1]. It is known as 2,3-butylene glycol, a valuable chemical feedstock because of its application as a solvent, a liquid fuel, and as a precursor of many synthetic polymers and resins. With a heating value of 27,200 J/g, 2,3-butanediol compares favorably with ethanol (29,100 J/g) and methanol (22,100 J/g) for use as a liquid fuel and fuel additive. Dehydration of 2,3-butanediol yields the industrial solvent methyl ethyl ketone. Further dehydration yields 1,3-butanediene, which is the starting material for synthetic rubber and is also an important monomer in the polymer industry. Methyl ethyl ketone can be hydrogenated to yield high octane isomers suitable for high quality aviation fuels. Diacetyl, formed by catalytic dehydrogenation of the diol, is a highly valued food additive. A wide variety of chemicals can also be easily prepared from 2,3-butanediol [2].

Interest in microbial production of 2,3-butanediol has been increasing recently due to the large number of industrial applications of this product [1]. Microbially produced 2,3-butanediol can be converted into 1,3-butadiene, a feedstock chemical currently supplied by the petrochemical industry. 1,3-butadiene can, in turn, be utilized in the manufacture of plastics, pharmaceuticals, and synthetic rubber [3]. Currently, the manufacturing of 2,3-butanediol is still growing by an annual rate of 4–7% due to the increased demand for polybutylene terephthalate resin, γ-butyrolactone, spandex, and their precursors [4].

The 2,3-butanediol production is dependent on various process variables [2, 4-6]. These studies demonstrated that optimization of media components and culture conditions are important for 2,3-butanediol production. The traditional method of optimization involves varying one factor at a time, while keeping the others constant. This strategy requires a relatively large number of experiments and frequently fails to anticipate the optimal conditions. This essential shortcoming is due to the inability of the approach to consider the effects of possible interactions between factors. The deficiency can be overcome by applying more efficient, statistically based experimental design. In this respect, Taguchi orthogonal design is important tools to determine the optimal process conditions. The advantages of using the Taguchi method are that many more factors can be screened and optimized simultaneously and much quantitative information can be extracted by only a few experimental trials. Therefore, these methods have been extensively applied in parameter optimization and process control [7].

Optimal operating parameters of 2,3-butanediol production using *Klebsiella pneumoniae* under submerged culture conditions are determined by using Taguchi method. The effect of different factors including medium composition, pH, temperature, mixing intensity, and inoculums size on 2,3-butanediol production was analyzed using the Taguchi method in three levels. Based on these analyses the optimum concentrations of acetic acid and succinic acid were found to be 0.5 and 1.0 (%w/v) respectively. Furthermore, optimum values for temperature, inoculum size, pH, and the shaking speed were determined as 37°C, 8 (g/l), 6.1, and 150 rpm respectively. The influence of the trace elements on 2,3-butanediol production was studied and it was shown that their addition to the medium was not necessary. Maximum 23-butanediol concentration under optimum conditions was obtained as 15.973 (g/l) after 48 hr of cell cultivation.

5.2 MATERIALS AND METHODS

5.2.1 Microorganism

Bacterial strain used in this study was *Klebsiella pneumoniae* PTCC 1290, obtained from the Iranian Research Organization for Science and Technology (IROST). The strain was maintained on nutrient agar slants at 4°C and subcultured monthly. The pre-culture medium was nutrient broth containing (per liter): 2.0 gm yeast extract, 5.0 gm peptone, 5.0 gm NaCl, and 1.0 gm beef extract, sterilized at 121°C for 15 min.

5.2.2 Taguchi Methodology

Taguchi method of design of experimental (DOE) involves establishment of large number of experimental situation described as orthogonal array (OA) to reduce experimental errors and to enhance their efficiency and reproducibility of the laboratory experiments [8]. The first step is to determine the various factors to be optimized in the culture medium that have critical effect on the 2,3-butanediol production. Factors were selected and the ranges were further assigned based on the group consensus consisting of design engineers, scientists and technicians with relevant experience.

Based on the obtained experimental data, six factors having significant influence on the 2,3-butanediol production were selected for the present Taguchi DOE study to optimize the submerged culture condition. Six factors (temperature, pH, agitation, inoculum size, acetic acid, and succinic acid) which showed significantly influence on the 2,3-butanediol production [1, 9, 10] were considered in the present experimental situation (Table 1).

TABLE 1 Selected fermentation factors and their assigned levels.

No.	Factor	Level 1	Level 2	Level 3
a	Temperature (°C)	28	32	37
b	pH	6.1	6.8	7.5
c	Agitation (rpm)	120	150	180
d	Inoculum size (g/L)	2	5	8
e	Acetic acid (%w/v)	0.1	0.5	1.0
f	Succinic acid (%w/v)	0.5	1.0	1.5

The next step was to design the matrix experiment and to define the data analysis procedure. The appropriate OAs for the control parameters to fit a specific study was selected. Taguchi et al. [11] provides many standard OAs and corresponding linear graphs for this purpose. In the present case, the three levels of factors variation were considered and the size of experimentation was represented by symbolic arrays L18 (which indicates eighteen experimental trails). Six factors with three levels were used and it is depicted in Table 1 and Table 2.

TABLE 2 Experimental setup (L18 Orthogonal Array).

Expt. No	Factor levels						2,3-butanediol production (g/L)
	a	b	c	d	e	f	
1	1	1	1	1	1	1	8.364
2	1	2	2	2	2	2	14.215
3	1	3	3	3	3	3	10.130
4	2	1	1	2	2	3	12.415
5	2	2	2	3	3	1	12.560
6	2	3	3	1	1	2	11.161
7	3	1	2	1	3	2	14.070

TABLE 2   *(Continued)*

Expt. No	Factor levels						2,3-butanediol production (g/L)
	a	b	c	d	e	f	
8	3	2	3	2	1	3	9.998
9	3	3	1	3	2	1	13.412
10	1	1	3	3	2	2	15.709
11	1	2	1	1	3	3	7.742
12	1	3	2	2	1	1	9.472
13	2	1	2	3	1	3	12.908
14	2	2	3	1	2	1	11.305
15	2	3	1	2	3	2	11.450
16	3	1	3	2	3	1	11.595
17	3	2	1	3	1	2	13.557
18	3	3	2	1	2	3	13.353

In the design OA, each column consists of a number of conditions depending on the levels assigned to each factor. Submerged fermentation experiments were carried out in cotton plugged 500 ml Erlenmeyer flasks containing 100 ml of production medium [(g/100 ml of distilled water) glucose: yeast extract 1: acetic acid (0.1, 0.5, and 1): succinic acid (0.5, 1.0, and 1.5): KH_2PO_4 0.15: K_2HPO_4 .3H2O 1.14: $(NH_4)_2SO_4$ 3: $MgSO_4$. $4H_2O$ .024: NaCl .01: EDTA .04: $CaCl_2.2H_2O$ 1.4×10^{-3}: $FeSO_4.7H_2O$ 1×10^{-3}: $ZnSO_4.7H_2O$ $.75\times10^{-3}$ and $MnSO_4.4H_2O$ 28×10^{-3} dissolved in 100 ml of distilled water and pH adjusted by adding NaOH or HCl prior to sterilization (15 min, 121°C). Glucose was sterilized separately].

Submerged fermentation experiments were performed for 2,3-butanediol production with *Klebsiella pneumoniae* PTCC 1290 employing selected eighteen experimental trails (Table 2) in combination with six factors at three levels (Table 1) and the result obtained from each set as 2,3-butanediol concentration (g/l) and was shown in Table 2.

5.2.3 Analysis

Cell concentration of the inoculum was determined by optical density measurement at 620 nm using a calibration curve to relate this parameter to cell mass dry weight.

The 2,3-butanediol concentrations were determined by a Fractovap 4,200 gas chromatograph (Carlo Erba, Milan, and Italy) using a Chromosorb 101 column (Supelco, Bellefonte, and PA) operated with N_2 as the carrier gas, at 250°C injector temperature, 300°C detector temperature, and 175°C column temperature, and using n-butanol as the internal standard.

5.2.4   Software

Qualitek-4 software (Nutek Inc., MI) for automatic design of experiments using Taguchi approach was used in the present study. Qualitek-4 software is equipped to use L4–L64 arrays along with selection of 2–63 factors with 1, 3, and 4 levels to each factor. The automatic design option allows Qualitek-4 to select the array used and assign factors to the appropriate columns. The obtained experimental data was processed in the Qualitek-4 software with bigger is better quality characteristics for the determination of the optimum culture conditions for the fermentation, to identify individual factors influence on the 2,3-butanediol production and to estimate the performance (fermentation) at the optimum conditions.

5.3   DISCUSSION AND RESULT

Submerged fermentation experiments studies with the designed experimental condition showed significant variation in the 2,3-butanediol production (Table 2). Production levels were found to be very much dependent on the culture conditions. The average affect of the factors along with interactions at the assigned levels on the 2,3-butanediol production by *Klebsiella pneumoniae* PTCC1290 was shown in Table 3.

TABLE 3   Main effects of the factors at the assigned levels on 2,3-butanediol production.

Factors	Level 1	Level 2	Level 3	L2–L1	L3–L2
Temperature	10.938	11.966	12.664	1.027	0.698
pH	12.510	11.562	11.496	-0.949	-0.660
Agitation	11.156	12.763	11.649	1.606	-1.115
Inoculum size	10.999	11.524	13.046	0.524	1.522
Acetic acid	10.910	13.401	11.257	2.490	-2.145
Succinic acid	11.118	13.360	11.090	2.241	-2.270

The difference between average value of each factor at higher level and lower level indicated the relative influence of the effect at their individual capacities. The positive or negative sign denoted variation of production values from level 1 to 2 or 3. Figure 1 shows the influence of each individual factor on the 2,3-butanediol production.

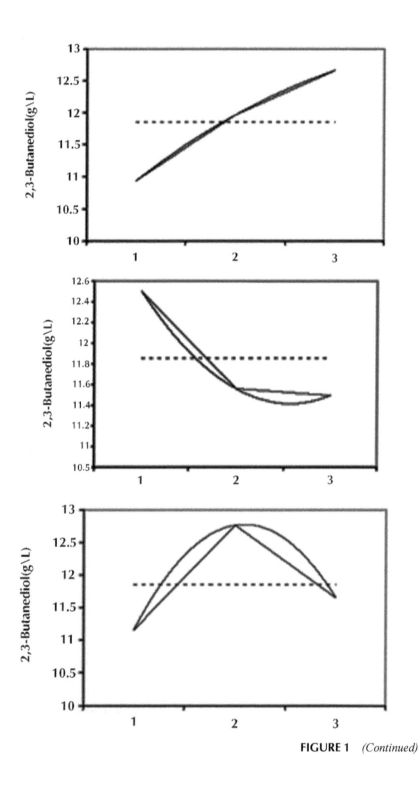

FIGURE 1   *(Continued)*

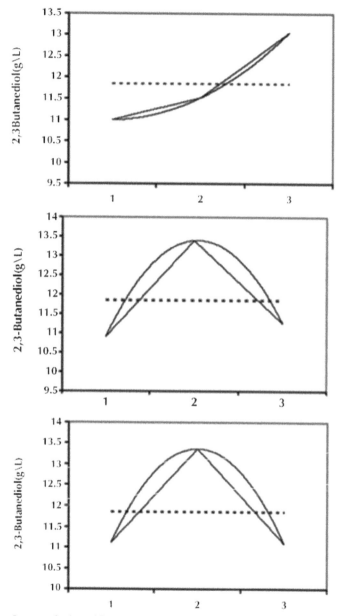

FIGURE 1   Impact of selected fermentation factor assigned level on 2, 3-butanediol production by *Klebsiella pneumoniae*. Impact of selected factor assigned levels on 2,3-butanediol production by *Klebsiella pneumoniae*. X-axis represents assigned levels of selected factor and Y-axis represents 2,3-butanediol production (g/l). (a) temperature, (b) pH, (c) agitation, (d) inoculum size, (e) acetic acid, and (f) succinic acid (…) indicates average 2,3-butanediol production during experimentation and (—) indicates individual factors contribution 2,3-butanediol production during experimentation.

Individually at level stage pH has highest affect in level 1 where as acetic acid and inoculum size has high affects in level 2 and 3 respectively on 2,3-butanediol concentration. It is clear that the primary factor affecting the substrate utilization rate in natural system is pH Syu [5] reported the effect of pH, on 2,3-butanediol production. He concluded that the maximum 2,3-butanediol formation was achieved at pH = 5.8 by *Fibrobacter succnogenes.* In the study the best pH was pH = 6.1.

The difference between level 2 and level 1 (L2−L1) of each factor indicates the relative influence of the affect. The larger the difference, the stronger is the influence. It can be seen from Table 3, that among the factors studied, acetic acid showed stronger influence compared to other factors followed by succinic acid, agitation, and temperature in the 2,3-butanediol production.

It is reported that 2,3-butanediol production can be increased by addition of different organic acid, because of they are intermediate metabolites for 2,3-butanediol production [12]. Nakashimada et al. [13] found that addition of acetate, propionate, pyruvate, and succinate enhanced 2,3-butanediol production. Among the organic acids giving an enhanced 2,3-butanediol concentration, acetate seemed to be the most appropriate additive because it gave the highest 2,3-butanediol production [13]. While acetate at high levels may be inhibitory to *Klebsiella pneumoniae*, low levels of acetate stimulate 2,3-butanediol production [12]. Stormer [14] noted that acetate in its ionized form induces acetolactate synthase formation, and thereby enhances the catalysis of pyruvate to 2,3-butanediol. The production of 2,3-butanediol by *Klebsiella oxytoca* NRRL B-199 was enhanced in the presence of low levels (>8 g/l) of lactate [15]. *Klebsiella oxytoca* ATCC 8724 grew well on xylose with 10 g/l succinate and produced additional 2,3-butanediol [16]. The production of 2,3-butanediol by *E. cloacae* NRRL B-23289 was also enhanced by the supplementation of acetate, lactate, and succinate [2]. New finding suggested that some amount of ethanol is formed by acetate reduction. Relative to this, a previous report demonstrated that acetate is converted to butanediol by condensation with pyruvate after the reduction of acetate to acetaldehyde [13]. In other work on cell-free extracts of *Aerobacter aerogenes* has demonstrated that acetate at low pH (i.e., in the form of acetic acid) serves as an effective inducer for the three enzymes involved in the formation of butanediol from pyruvate, *viz.*, pH 6 acetolactate-forming enzyme, acetolactate decarboxylase and diacetyl (acetoin) reductase [12]. Such an induction mechanism probably plays a major role in the enhanced butanediol production of our study, even though the exact extent of stimulation is not known. Our finding confirms increasing effect of acetic acid on 2,3-butanediol production. In the study 2,3-butanediol production of *Klebsiella pneumoniae* at initial substrate concentrations was considerably enhanced by the addition of 0.5% (w/v) acetic acid to the media.

Increasing of temperature and inoculum size has resulted in increase 2,3-butanediol production. Perego et al. [1] in an optimization study on 2,3-butanediol production by *B.licheniformis* (NCIMB 8059) found that butanediol concentration have a progressive increasing, when temperature was increased from 34 to 37°C. Conversely, they all sharply decreased over 37°C, likely due to the well-known thermal inactivation of biosystems at temperature higher than the optimum. Thus, supporting the assumption of considering 2,3-butanediol production as a process controlled enzymatically. On the other hand carbon consumption depends on the culture temperature [6].

The inoculum size was reported to improve the rate of 2,3-butanediol formation but not its yield on consumed carbon source. An optimization study of glucose fermentation by *B. licheniformis*, likely performed using a factorial experimental design demonstrated that an increase in the inoculum size had positive effect on the yield as well [17].

Agitation is another important factor for 2,3-butanediol production. Saha et al. [2] postulate that aeration may be of value in removing carbon dioxide produced in the process and thus have a stimulatory effect on the fermentation. These results further confirmed that, each studied factor was important in 2,3-butanediol production, and the influence of one factor on 2,3-butanediol production was dependent on the condition of the other factor in optimization of 2,3-butanediol production, although they have different influence at their individual levels. Although 2,3-butanediol is a product of anaerobic fermentation: aeration is known to enhance its production [18]. In the case of agitation increase to level 2 resulted in increase and subsequent increase to level 3, showed decrease in 2,3-butanediol concentration. This may be reasoned due to the other constitutive effect of culture media. Increasing pH has reverse effect in 2,3-butanediol production (Figure 1 and Table 3).

Understanding the interaction between two factors gives a better insight into the overall process analysis. Any individual factor may interact with any or all of the other factors creating the possibility of presence of a large number of interactions. This kind of interaction is possible in Taguchi DOE. Estimated interaction severity index (SI) of the factors under study helps to know the influence of two individual factors at various levels of the interactions (Table 4). In the Table, the "columns" represent the locations to which the interacting factors are assigned. Interaction SI presents 100% of SI for 90° angle between the lines while, 0% SI for parallel lines. "Reserved column" shows the column that should be reserved if this interaction effect has to be studied. "Levels" indicate the factor levels desirable for the optimum conditions (based on the first two levels).

TABLE 4   Estimated interaction of severity index for different parameters.

Interacting factors	column	SI (%)	Reserved col	Level
Temperature* Succinic acid	(a, f)	55.18	5	(1,2)
Temperature* Acetic acid	(a, e)	51.44	4	(1,2)
Agitation* Inoculum	(c, d)	50.78	1	(2,1)
pH* Acetic acid	(b, e)	31.23	5	(1,2)
Temperature* Inoculum	(a, d)	27.31	7	(3,1)
Temperature* Agitation	(a, c)	26.41	6	(3,2)

TABLE 4 *(Continued)*

Interacting factors	column	SI (%)	Reserved col	Level
pH* Succinic acid	(b, f)	24.55	4	(1,2)
pH* Inoculum	(b, d)	18.57	16	(1,3)
Agitation* Succinic acid	(c, f)	18.52	3	(2,2)
Inoculum* Acetic acid	(d, c)	10.50	3	(3,2)
Agitation* Acetic acid	(c, e)	7.65	2	(2,2)
Acetic acid* Succinic acid	(e, f)	6.92	14	(2,2)
pH* Agitation	(b, c)	5.54	7	(1,3)
Inoculum* Succinic acid	(d, f)	5.02	2	(3,2)
Temperature* pH	(a, b)	2.59	1	(3,3)

In the study, interaction between two selected factors has shown in Table 4. The interaction was measured based on SI value calculated by software program. This value between two selected factors varied (2.6-55%) with factor-to-factor (Table 4). From the Table it can be followed that temperature and succinate (at level 1 and 2, column 5) interactions showed highest interaction SI (55.18%) followed by temperature and acetate (at level 1 and 2, column 4) with 51.44%.

In Taguchi approach, analysis of variance (ANOVA) is used to analyze the results of the OA experiment and to determine how much variation each factor has contributed. From the calculated ratios (F), it can be referred that all factors and interactions considered in the experimental design are statically significant effects at 95% confidence limit, indicating that the variability of experimental data explained in terms of significant effects. By studying the main effects of each of the factors, the general trends of the influence of the factors towards the process can be characterized. The characteristics can be controlled such that a lower or a higher value in a particular influencing factor produces the preferred result. Thus, the levels of factors, to produce the best results can be predicted. The ANOVA with the percentage of contribution of each factor with interactions are shown in Table 5. It can be observed from the table that acetic acid is the most significant factor for the 2,3-butanediol production. Succinic acid and inoculum size are the next most important significant factors in the 2,3-butanediol production. The pH showed least impact among the factors studied with the assigned variance of values. The error observed was very low which indicated the accuracy of the experimentation (Figure 2).

TABLE 5 Analysis of variance (ANOVA).

Factors	DOF	sum of squares (S)	variance (V)	F -ratio (F)	pure sum (S')	precent (P%)
Temperature	2	9.041	4.520	93.406	8.944	11.610
pH	2	3.859	1.929	39.873	3.672	4.884
Agitation	2	8.125	4.062	83.950	8.029	10.422
Inoculum size	2	13.562	6.781	140.118	13.465	17.479
Acetic acid	2	21.847	10.923	225.717	21.751	28.235
Succinic acid	2	20.353	10.178	210.315	20.260	26.299
Other/error	5	0.541	0.048			1.071
Total	17	77.035				100.00

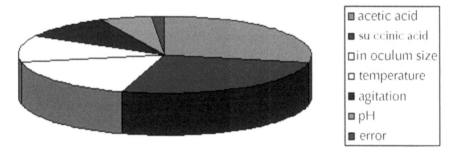

■ acetic acid
■ su ccinic acid
□ in oculum size
□ temperature
■ agitation
■ pH
■ error

FIGURE 2 Relative influence of factors and interaction.

Table 6 represents the optimum conditions required for the production of maximum 2,3-butanediol by this bacterial strain. Based on software prediction, the average performance of this strain in 2,3-butanediol production was observed to be 11.856 (g/l). The data also suggested that organic acids play a vital role contributing 46.15% in butanediol production under the optimized conditions.

The 2,3-butanediol production can be increased from 11.856 to 18.459 (g/l) that is overall 35.8% enhancement in the production can be achieved. Further to validate the proposed experimental methodology, fermentation experiments were performed for 2,3-butanediol production by employing the obtained optimized culture conditions (Table 6). The experimental data showed an enhanced 2,3-butanediol concentration of 15.973 (g/l) from 11.856 (g/l) (28.3% improvement in butanediol production) with the modified culture conditions.

TABLE 6   Optimal conditions and their performance in production of 2,3-butanediol.

Factors	Level description	Level	Contribution
Temperature	37	3	0.807
pH	6.1	1	0.653
Agitation	150	2	0.906
Inoculum size	8	3	1.189
Acetic acid	0.5	2	1.544
Succinic acid	1.0	2	1.503

Total contribution from all factors	6.603
Current grand average performance	11.856
Expected result at optimal conditions	18.459

The study of interactive influence of selected factors (Table 6) revealed a unique relationship such as showing low influence on product production at individual level and higher SI at interactive level (Table 4), indicating the importance of parameter optimization on any product production and the role of various physic-chemical parameters including organic acids concentration, agitation, temperature and pH of the medium in microbial metabolism. Such factor mediated regulation of microbial fermentation was observed with many microbial species on any product [19].

5.4   CONCLUSION

Culture conditions and media composition optimization by a conventional one-at-the-approach led to a substantial increase in 2,3-butanediol concentration. However, this approach is not only cumbersome and time consuming, but also has the limitation of ignoring the importance of interaction of various parameters. Taguchi approach of OA experimental design for process optimization, involving a study of given system by a set of independent variables (factors) over a specific region of interest (levels) by identifying the influence of individual factors, establish the relationship between variables and operational conditions and finally establish the performance at the optimum levels obtained. In this methodology, the desired design is sought by selecting the best performance under conditions that produces consistent performance leads to a more fully developed process. The obtained optimal culture condition for the 2,3-butanediol production from the proposed methodology was validated by performance the experiments with the obtained conditions.

KEYWORDS

- **Analysis of variance**
- **Cell cultivation**
- **Design of experimental**
- **Microorganism**
- **Orthogonal array**

REFERENCES

1. Perego, P., Converti, A., and Del Borghi, M. Effects of temperature, inoculum size, and starch hydrolyzate concentration on butanediol production by *Bacillus licheniformis*. *Bioresource Technology*, **89**, 125–131 (2003).

2. Saha, B. C. and Bothast, R. J. Production of 2,3-butanediol by newly isolated *Enterobacter cloacae*. *Appl. Microbiol. Biotechnol.*, **52**, 321–326 (1999).

3. Mallonee, D. H. and Speckman, R. A. Development of a Mutant Strain of *Bacillus polymyxa* Showing Enhanced Production of 2,3-butanediol. *Applied and Environmental Microbiology*, **45**(1), 168–171 (1988).

4. Jiayang, Q., Zijun, X., and Cuiqing, M. Production of 2,3-butanediol by *Klebsiella pneumoniae* using glucose and ammonium phosphate. *Chinese J. Chem. Eng.*, **14**(1), 132–136 (2006).

5. Syu, M. J. Biological production 2,3-butanediol. *Appl. Microbiol. Biotechnol.*, **55**, 10–18 (2001).

6. Marwoto, B., Nakashimada, Y., Kakizono, T., and Nishio, N. Enhancement of *(R,R)*-2,3-butanediol production from xylose by *Paenibacillus polymyxa* at elevated temperatures. *Biotechnology Letters*, **24**, 109–114 (2002).

7. Hao, D. C., Zhu, P. H., and Yang, S. L. Optimization of recombinant Cytochrome P450 2C9 protein production in *Escherichia coli* DH5a by statistically-based experimental design. *World J. Microbiol. Biotechnol.*, **22**, 1169–1176 (2006).

8. Montgomery, D. C. *Design and analysis of experiments*. John Wiley & Sons, New York ISBN 0-471-48735-X (2004).

9. Perego, P., Converti, A., Del Borghi, A., and Canepa, P. 2,3-butanediol production by *Enterobacter aerogenes*: Selection of optimal condition and application to food industry residues. *Bioprocess Engineering*, **23**, 613–620 (2000).

10. Ghosh, S. and Swaminathan, T. Optimization of Process Variables for the Extractive Fermentation of 2,3-butanediol by *Klebsiella oxytoca* in Aqueous Two-phase System Using Response Surface Methodology. *Chem. Biochem. Eng. Q.* **17**(4), 319–325 (2003).

11. Taguchi, G., Chowdhury, S., and Wu, Y. *Taguchi's quality engineering handbook*. John Wiley & Sons, New York ISBN 0-471-41334-8 (2004).

12. Yu, E. K. C. and Saddler, J. N. Enhanced Production of 2,3-butanediol by *Klebsiella pneumoniae* Grown on High Sugar Concentrations in the Presence of Acetic Acid. *Applied and Environmental Microbiology*, **44**(4), 784–777 (1982).

13. Nakashimada, Y., Marwoto, B., Kashiwamura, T., Kakizono, T., and Nishio, N. Enhanced 2,3-butanediol Production by Addition of Acetic Acid in *Paenibacillus polymyxa*. *J. of Bioscience and Bioengineering*, **90**(6), 661–664 (2000).

14. Stormer, F. C. Evidence for regulation of *Aerobacter aerogenes* pH 6 acetolactate forming enzyme by acetate ion. *Biochem Biophys Res Commun*, **74**, 898–902 (1977).

15. Qureshi, N. and Cheryan, M. Effect of lactic acid on growth and butanediol production by *Klebsiella oxytoca*. *J Ind Microbiol*, **4**, 453–456 (1989).

16. Eiteman, M. A. and Miller, J. H. Effect of succinic acid on 2,3-butanediol production by *Klebsiella oxytoca*. *Biotechnol Letts.*, **17**, 1057–1062 (1995).

17. Nilegaonkar, S., Bhosale, S. B., Dandage, C. N., and Kapidi, A. H. Potential of *Bacillus licheniformis* for the production of 2,3-butanediol. *J. Ferment. Bioeng.*, **82**, 408–410 (1996).

18. Jansen, N. B., Flickinger, M. C, and Tsao, G. T. Production of 2,3-butanediol from xylose by *Klebsiella oxytoca* ATCC 8724. *Biotechnol Bioeng.*, **26**, 362–368 (1984).

19. Prakasham, R. S., Subba Rao, Ch., Sreenivas Rao, R., and Sarma, P. N. Enhancement of acid amylase production by an isolated *Aspergillus awamori*. *Journal of Applied Microbiology*, **102**, 204–211 (2007).

6 New Trends and Achievements on Solvent Extraction of Copper

CONTENTS

6.1 INTRODUCTION

Pure Cu(I) is soft and malleable, an exposed surface has a reddish-orange tarnish. It is used as a conductor of heat and electricity, a building material, metal processing, and metal finishing [1, 2]. Cu(II) ions are water-soluble, where they function at low concentration as bacteriostatic substances, fungicides, and wood preservatives. In sufficient amounts, they are poisonous to higher organisms, at lower concentrations it is an essential trace nutrient to all higher plant and animal life [3].

Various techniques remove Cu(I) from aqueous solutions, membrane filtration, flotation, electrolysis, biosorption, precipitation [4, 5] and liquid-liquid extraction. Between various techniques extract Cu(I) from aqueous solutions, liquid-liquid extraction is one of the effective techniques to extract Cu(I) from aqueous solutions [6]. So, liquid-liquid extraction is established technologic for recovery of metals from dilute aqueous product effluents [7-12]. In this chapter, solvent extraction of Copper (I) {Cu(I)} from aqueous solutions by organic solvent composing of lauric acid (extractant)/benzene (diluent) has been studied at T = 298.2K under atmospheric pressure. Effect of initial Cu(I) concentration in aqueous phase (5×10^{-4}, 2.5×10^{-3}, 5×10^{-3}, and 2.5×10^{-2}) M in pH = 1.6 sulfuric acid solution were investigated. Extraction was studied as a function of organic phase composition, acid concentration, aqueous pH, primary copper concentration. The copper concentrations were analyzed by spectroscopy. Percentage extraction (%E) of Cu(I) was studied. Distribution coefficient (D') were measured to determine the extracting capability of the extractant. The (D') and (%E) increased with growth of pH. The results indicate pH, (%E), and (D') decrease with increasing initial Cu(I) concentration (5×10^{-4} to 2.5×10^{-2}) M and they decrease with increase of lauric acid concentration (0.2–0.5) m.

6.2   EXPERIMENTAL

6 2.1   Materials

All the chemicals were used without further purification. Copper sulfate pentahydrate ($CuSO_4.5H_2O$) (Merck $\geq$ 99.6% purity), benzene (Merck $\geq$ 99% purity), lauric acid (probus $\geq$ 99% purity) sulfuric acid (H_2SO_4) (Probus R.A $\geq$ 98% purity) were purchased. Distilled and deionized water was used throughout all experiments. The structure of lauric acid is shown in I.

I

6.2.2   Extraction Procedures

A volume of 10 ml of Cu(II) were prepared by dissolving appropriate amounts of $CuSO_4.5H_2O$ (5×10^{-4}, 2.5×10^{-3}, 5×10^{-3}, and 2.5×10^{-2}) M in distilled water loaded with 0.1M Na_2SO_4 and containing aqueous phase at 1:1 organic to aqueous volume ratio in glass cell. Organic phase was prepared with lauric acid (extractant) and benzene (diluents). A glass cell connected to a water thermostat was made to measure the liquid-liquid extraction data. The prepared mixtures were introduced into the extraction cell and were stirred for 2 hr, and then left to settle for 2 hr for phase separation. After being allowed to reach equilibrium, samples were carefully taken from each phase. Aqueous phase pH was measured with a radiometer Copenhagen pH meter (model 62). The samples was used to 1.8 g/dm³ H_2SO_4, pH = 1.6. The concentration of the Cu(I) in organic phase was obtained from a Hitachi UV–V is Spectrophotometer (model 40–100) at a wavelength of 412 nm. All the experiments were carried out at constant temperature at T = 298.2K. The temperature was estimated to be accurate to within solution ±0.1 K. The (%E) of Cu(I) were calculated according to [7, 8]:

$$\% \text{ E} = \frac{[Cu]_{initial,aq} - [Cu]_{r,aq}}{[Cu]_{initial,aq}} \times 100 \tag{1}$$

where $[Cu]_{initial,aq}$ is the initial Cu(I) concentration in the aqueous phase and $[Cu]_{r,aq}$ is the remaining Cu(I) concentration in the aqueous phase after extraction.

6.3 THEORY

The extraction process may be represented by the equation:

$$2Cu^{2+} + 3\,(\overline{HR})_2 \Leftrightarrow \overline{(CuR_2.HR)_2} + 4H^+ \tag{2}$$

where $(\overline{HR})_2$ represents the extraction reagent. The extraction constant of the species $\overline{(CuR_2.HR)_2}$ is given by:

$$K_{ex} = \frac{\left[\overline{CuR_2.\,HR_2}\right]\left[H^+\right]^4}{\left[Cu^{2+}\right]^2\left[\overline{HR_2}\right]^3} \tag{3}$$

where $\overline{(CuR_2.HR)_2}$ is the only extractable species. Introducing the mass balance equation for the Cu(I) and the extractant:

$$\left[\overline{Cu}\right] = 2\left[\overline{(CuR_2.HR)_2}\right] \tag{4}$$

The mass balance Equation for the Cu(I) in the aqueous phase:

$$[Cu] = [Cu^{2+}] + [CuR^+] + [CuR_2] + [CuSO_4] \tag{5}$$

$$[Cu] = \alpha_{Cu}\,[Cu^{2+}] \tag{6}$$

where α_{Cu} is formation constant of the Cu(I). The extraction constant of the total Copper in the aqueous phase is given by:

$$k'_{ex} = \frac{\left[\overline{CuR_2.\,HR_2}\right]\left[H^+\right]^4}{\left[Cu^{2+}\right]^2\left[\overline{H_2R_2}\right]^3} \tag{7}$$

The metal distribution ratio (D') and the extraction constant is related by:

$$D' = \frac{\left[\overline{Cu}\right]}{[Cu]} = 2\,k'_{ex}\,[Cu]\,[\,\overline{H_2R_2}\,]^3\,[H^+]^4 \tag{8}$$

Therefore, $\log D' = 0.3010 + \log k_{ex}' + \log [Cu] + 3 \log [\overline{H_2R_2}] + 4pH$ (9)

According to Equation (9) a plot of log D' *versus* pH will give straight line shown in Figure 1 and Figure 2. The distribution coefficients (D') increase with increasing pH. A plot of log D' – log [Cu] against pH (Figure 3) will give a straight line of slope and intercept log K $_{ex}$ at constant concentration of log $[\overline{H_2R_2}]$ in Table 1. The Figure 4 (log D' – log [Cu] – 4pH *versus* of log $[\overline{H_2R_2}]$) will show a straight line of slope and intercept log K$_{ex}$. Parameter correlation equilibrium data for extraction of Cu(I) with using lauric acid diluted in benzene at 298.2K presented in Table 2. The extraction constant (log K$_{ex}$) has been calculated as –12.2580.

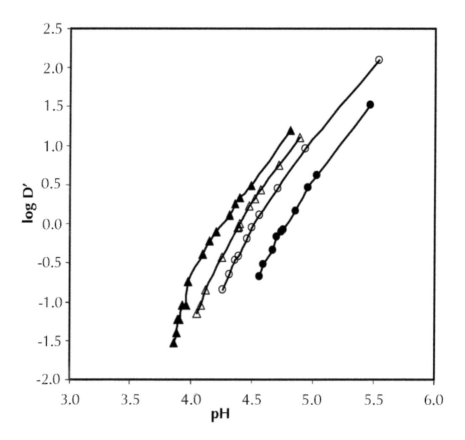

FIGURE 1 The plot of distribution coefficient (log D') against pH at the extraction of Cu (I) with lauric acid (0.2 M) diluted in benzene at 298.2 K. Initial concentration of Cu(I) (M): (●) 5×10^{-4}; (○) 2.5×10^{-3}; (Δ) 5×10^{-3}; (▲) 2.5×10^{-2}.

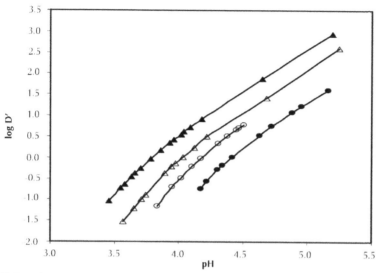

FIGURE 2 The plot of distribution coefficient (log *D*') against pH at the extraction of Cu(I) with lauric acid (0.5 M) in benzene at 298.2 K. Initial concentration of Cu(I) (M): (●) 5×10^{-4}; (○) 2.5×10^{-3}; (△) 5×10^{-3}; (▲) 2.5×10^{-2}.

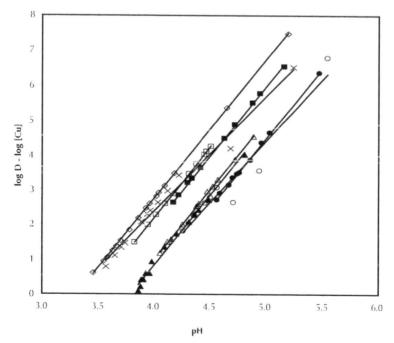

FIGURE 3 The plots of log *D*' –log [Cu] versus pH at the extraction of Cu(I) with lauric acid diluted in benzene at 298.2 K. Initial concentration of Cu(I) (M): concentration of 0.2 M lauric acid: (●) 5×10^{-4}; (○) 2.5×10^{-3}; (△) 5×10^{-3}; (▲) 2.5×10^{-2} and concentration of 0.5 M lauric acid: (■) 5×10^{-4}; (□) 2.5×10^{-3}; (×) 5×10^{-3}; (◊) 2.5×10^{-2}.

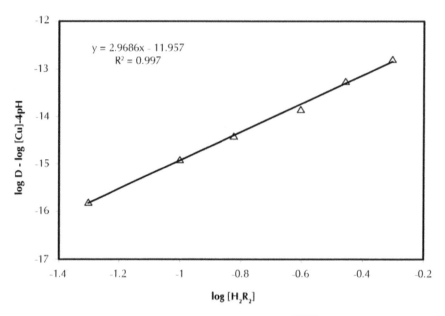

FIGURE 4   The plots of log D' – log [Cu] – 4pH against log [$\overline{H_2R_2}$] of the extraction of Cu(I) with lauric acid diluted in benzene at 298.2 K. Initial concentration of Cu(I) (M): (●) 5×10⁻⁴; (○) 2.5×10⁻³; (Δ) 5×10⁻³; (▲) 2.5×10⁻².

TABLE 1   Equilibrium data of (CuSO₄ + lauric acid + benzene) at 298.2K: influence of luaric acid concentration. $C_{Cu,initial}$ = 5×10⁻³ M , $C_{Na_2So_4}$ = 0.1 M (14.204 g/dm³).

$\overline{Cu} \times 10^{-3}$ (mol-g/dm³)	$[Cu] \times 10^{-3}$ (mol-g/dm³)	D'	pH	$\overline{[H_2R_2]}$ (mol-g/dm³)
1.88	3.12	0.60	4.52	0.05
1.89	3.11	0.61	4.30	0.10
1.91	3.09	0.62	4.18	0.15
1.99	3.01	0.66	4.05	0.25
2.05	2.95	0.70	3.91	0.35
2.09	2.91	0.72	3.80	0.50

TABLE 2 Parameter correlation equilibrium data for extraction of Cu(I) with using lauric acid diluted in benzene at 298.2K.

C_{acid} = 0.2 M (40.07 g/dm³)					C_{acid} = 0.5 M (100.17 g/dm³)				
$[\overline{Cu}]\times10^{-3}$ (mol-g/dm³)	$[Cu]\times10^{-3}$ (mol-g/dm³)	D'	%E	pH	$[\overline{Cu}]\times10^{-3}$ (mol-g/dm³)	$[Cu]\times10^{-3}$ (mol-g/dm³)	D'	%E	pH
0.310	2.19	0.14	12.4	4.26	0.171	2.33	0.07	6.8	3.83
0.454	2.05	0.22	18.2	4.31	0.419	2.08	0.20	16.8	3.95
0.638	1.86	0.34	25.5	4.36	0.790	1.71	0.32	31.6	4.02
0.691	1.81	0.38	27.6	4.39	0.945	1.56	0.61	37.8	4.10
0.974	1.53	0.64	39.0	4.46	0.122	1.28	0.96	48.9	4.17
1.03	1.47	0.89	41.2	4.50	0.172	0.784	2.19	68.7	4.31
1.40	1.10	1.28	56.1	4.56	0.191	0.587	3.26	76.5	4.38
1.85	6.52	2.83	73.9	4.71	2.05	0.447	4.60	82.1	4.45
2.25	2.52	8.91	89.9	4.94	2.09	0.412	5.07	83.5	4.47
2.48	0.02	123	99.2	5.54	2.15	0.353	6.08	85.9	4.51

6.4 DISCUSSION AND RESULTS

Data extraction equilibrium for initial Cu(I) concentration (5×10^{-4}, 2.5×10^{-3}, 5×10^{-3}, and 2.5×10^{-2}) M in pH = 1.6 sulfuric acid solution in the range of 0.2–0.5 M lauric acid, pH for extraction, (D'), (%E) given in Tables 3–6 at atmospheric pressure and at T = 298.2 K. Table 1 gives influence of lauric acid concentration of the equilibrium data of $CuSO_4$ and lauric acid diluted in benzene.

TABLE 3   Equilibrium data of ($CuSO_4$ + lauric acid + benzene) at 298.2K. $C_{Cu,initial} = 5 \times 10^{-4}$ M (0.0318 g/dm³), $C_{Na_2SO4} = 0.1$ M (14.204 g/dm³).

$C_{Cu, initial}$ (mol-g/dm³)	Log D'– log [Cu] versus pH					
	$C_{acid} = 0.2$ M			$C_{acid} = 0.5$ M		
	a	b	R²	a	b	R²
5.0×10^{-4}	3.9884	−15.460	0.9984	3.971	−13.907	0.9997
2.5×10^{-3}	3.5625	−13.412	0.9294	4.0632	−14.070	0.9999
5.0×10^{-3}	3.9939	−14.976	0.9996	3.5488	−11.741	0.9906
2.5×10^{-2}	4.0035	−15.180	0.9907	3.9927	−13.227	0.9999

6.4.1 Effect of pH and Extractant Concentration

Creature Cu(I) extraction by lauric acid depend on the initial acidity of the aqueous solution, more studied were carried out in the order to recognize the influence of the aqueous pH on Cu(I) extraction. The pH will be very essential factor in the separation of metal ion. The results obtained were shown in Figure 5 and Figure 6, plotting the (%E) against equilibrium pH. From Figure 5 and Figure 6 find the lowest at pH of 3.46 and maximum pH of 5.54. The (%E) increased with growth of pH. It also can be seen that the curves are shifted to the left as the increase of initial Cu(I) concentration.

6.4.2 Thermodynamic Part

Free energy ($\Delta G°$) (thermodynamic parameter) concerned in this study due to the transfer of a unit mole of Cu(I) from the aqueous into the organic phases. The $\Delta G°$ of Cu(I) extraction at T = 298.2K was determined from [13].

$$\Delta G° = -2.303 RT \log K_{ex} \tag{10}$$

where R is the universal gas constant (8.314 J/mol K), T is the thermodynamic temperature and K_{ex} is the extraction constant and this was presented in Table 7 and the $\Delta G°$ value is negative. The negative of $\Delta G°$ indicated that the Cu(I) extraction with lauric acid occurred spontaneously at T = 298.2K. From log K_{ex} versus log [Cu] (Figure 7) the extraction constant (log K_{ex}) increased with increasing Cu(I) concentration in aqueous phase.

TABLE 4 Equilibrium data of (CuSO$_4$ + lauric acid + benzene) at 298.2K. $C_{Cu,initial} = 2.5\times10^{-3}$ M (0.1589 g/dm³), $C_{Na_2SO_4} = 0.1$ M (14.204 g/dm³).

$C_{acid} = 0.2$ M (40.07 g/dm³)					$C_{acid} = 0.5$ M (100.17 g/dm³)				
$[\overline{Cu}]\times10^{-4}$ (mol-g/dm³)	$[Cu]\times10^{-4}$ (mol-g/dm³)	D'	%E	pH	$[\overline{Cu}]\times10^{-4}$ (mol-g/dm³)	$[Cu]\times10^{-3}$ (mol-g/dm³)	D'	%E	pH
0.866	4.14	0.21	17.3	4.56	0.775	4.23	0.18	15.5	4.17
1.15	3.86	0.30	22.9	4.59	1.07	3.93	0.27	21.5	4.22
1.58	3.42	0.46	31.6	4.67	1.70	3.30	0.52	34.1	4.30
2.01	2.99	0.67	40.1	4.70	1.98	3.06	0.65	39.6	4.34
2.21	2.80	0.79	44.1	4.74	2.55	2.45	1.04	51.0	4.42
2.27	2.73	0.83	45.4	4.76	3.87	1.13	3.43	77.4	4.63
2.96	2.04	1.46	59.3	4.86	4.24	0.759	5.59	84.8	4.72
3.72	1.28	2.90	74.4	4.96	4.62	0.386	12.0	92.3	4.88
4.03	0.975	4.13	80.5	5.03	4.72	0.284	16.6	94.3	4.95
4.86	0.147	33.1	97.1	5.47	4.88	0.119	40.9	97.6	5.16

TABLE 5 Equilibrium data of ($CuSO_4$ + lauric acid + benzene) at 298.2K. $C_{Cu,initial}$ = 5×10^{-3} M (0.3177 g/dm³), $C_{Na_2SO_4}$ = 0.1 M (14.204 g/dm³).

C_{acid} = 0.2 M (40.07 g/dm³)					C_{acid} = 0.5 M (100.17 g/dm³)				
$[Cu]\times10^{-3}$ (mol/dm³)	$[Cu]\times10^{-3}$ (mol/dm³)	D'	$\%E$	pH	$[Cu]\times10^{-3}$ (mol/dm³)	$[Cu]\times10^{-3}$ (mol/dm³)	D'	$\%E$	pH
0.309	4.69	0.07	6.2	4.05	0.148	4.85	0.03	3.0	3.57
0.422	4.58	0.09	8.4	4.08	0.293	4.71	0.06	5.9	3.65
0.603	4.40	0.14	12.1	4.12	0.463	4.54	0.10	9.3	3.71
1.35	3.65	0.37	27.0	4.25	0.536	4.47	0.13	11.3	3.74
2.35	2.66	0.88	47.0	4.38	1.47	3.53	0.42	29.4	3.89
2.40	2.60	0.92	48.0	4.39	1.89	3.11	0.61	37.8	3.94
2.49	2.51	0.99	49.8	4.40	2.10	2.91	0.72	42.0	3.97
3.12	1.88	1.66	62.4	4.48	2.55	2.46	1.04	51.2	4.03
3.34	1.66	2.04	67.2	4.52	3.15	1.85	1.70	63.0	4.12
3.64	1.36	2.68	72.8	4.57	3.80	1.21	3.14	76.0	4.22
4.24	0.762	5.56	84.8	4.72	4.83	1.69	25.6	96.7	4.68
4.63	0.370	12.5	92.6	4.89	4.99	1.22	396	99.7	5.24
C_{acid} = 0.2 M(40.07 g/dm³)					C_{acid} = 0.5 M(100.17 g/dm³)				

TABLE 6 Equilibrium data of (CuSO$_4$ + lauric acid + benzene) at 298.2K. $C_{Cu,initial}$ = 2.5×10^{-2} M (1.5886 g/dm^3), $C_{Na_2SO_4}$ = 0.1 M (14.204 g/dm^3).

C$_{acid}$ = 0.2 M(40.07 g/dm^3)					C$_{acid}$ = 0.5 M(100.17 g/dm^3)				
$[Cu]$×10^{-2} (mol/dm^3)	$[\overline{Cu}]$×10^{-2} (mol/dm^3)	D'	%E	pH	$[Cu]$×10^{-2} (mol/dm^3)	$[\overline{Cu}]$×10^{-2} (mol/dm^3)	D'	%E	pH
0.067	2.43	0.03	2.7	3.86	0.205	2.23	0.09	8.2	3.46
0.10	2.40	0.04	4.0	3.88	0.393	2.11	0.19	15.7	3.55
0.132	2.37	0.06	5.3	3.89	0.467	2.03	0.23	18.7	3.58
0.136	2.36	0.06	5.4	3.90	0.629	1.87	0.34	25.2	3.63
0.150	2.35	0.06	6.0	3.91	0.735	1.77	0.42	29.4	3.66
0.199	2.30	0.09	8.0	3.93	0.889	1.61	0.55	35.5	3.70
0.216	2.28	0.09	8.6	3.96	1.20	1.30	0.92	48.0	3.78
0.381	2.12	0.18	15.2	3.98	1.51	0.996	1.51	60.2	3.86
0.724	1.78	0.41	29.0	4.09	1.74	0.763	2.28	69.5	3.93
0.934	1.57	0.60	37.4	4.15	1.82	0.676	2.70	73.0	3.96
1.10	1.41	0.78	43.8	4.20	1.95	0.551	3.54	78.0	4.02
1.41	1.09	1.29	56.4	4.31	2.01	0.493	4.08	80.4	4.04

TABLE 6 *(Continued)*

$C_{acid} = 0.2$ M(40.07 g/dm³)					$C_{acid} = 0.5$ M(100.17 g/dm³)				
$[\overline{Cu}]\times10^{-2}$ (mol/dm³)	$[\overline{Cu}]\times10^{-2}$ (mol/dm³)	D'	%E	pH	$[\overline{Cu}]\times10^{-2}$ (mol/dm³)	$[\overline{Cu}]\times10^{-2}$ (mol/dm³)	D'	%E	pH
1.59	0.908	1.80	63.7	4.36	2.10	0.401	5.24	84.0	4.09
1.69	0.807	2.10	67.6	4.40	2.23	0.274	8.12	89.0	4.18
1.89	0.614	3.07	75.4	4.49	2.47	0.033		73.9	98.7	4.65
2.35	0.152	15.4	93.9	4.81	0.25	0.003		874	99.5	5.19

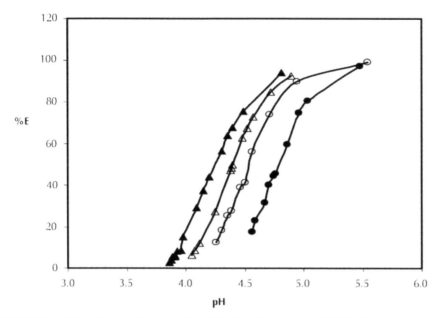

FIGURE 5 Effect of pH on the extraction of Cu (I) with lauric acid (0.2 M) diluted in benzene at 298.2 K. Initial concentration of Cu (I) (M): (●) 5×10^{-4}; (○) 2.5×10^{-3}; (Δ) 5×10^{-3}; (▲) 2.5×10^{-2}.

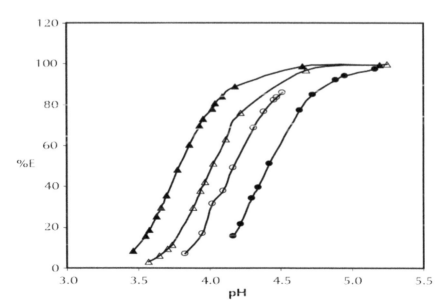

FIGURE 6 Effect of pH on the extraction of Cu (I) with lauric acid (0.5 M) diluted in benzene at 298.2 K. Initial concentration of Cu (×) (M): (●) 5×10^{-4}; (○) 2.5×10^{-3}; (Δ) 5×10^{-3}; (▲) 2.5×10^{-2}.

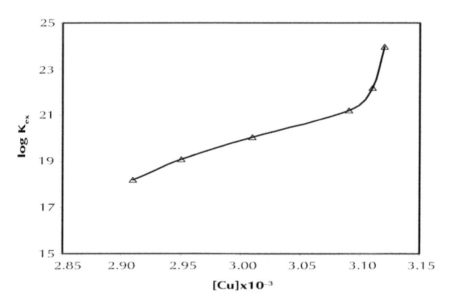

FIGURE 7 The plot of log K_{ex} against [Cu] of the extraction of Copper (I) with lauric acid diluted in benzene at 298.2 K: Initial concentration of Cu(I) (M): (Δ) 5×10^{-3}.

TABLE 7 The extraction constant and free energy ($\Delta G°$) of equilibrium data for extraction of Cu (I) with using lauric acid diluted in benzene at 298.2K.

[Cu] $\times 10^{-3}$ (mol-g/dm³)	K_{ex}	$\Delta G°$ (kJ/mol)
3.12	$9.248 \times 10^{+23}$	−136.838
3.11	$1.554 \times 10^{+22}$	−126.707
3.09	$1.560 \times 10^{+21}$	−121.006
3.01	$1.112 \times 10^{+20}$	−114.457
2.95	$1.208 \times 10^{+19}$	−108.952
2.91	$1.568 \times 10^{+18}$	−103.890

6.5   CONCLUSION

The experimental data indicate lauric acid in organic phase extracts capably Cu(I) from aqueous solutions in the pH = 1.6 Sulfuric acid at T = 298.2K under atmospheric pressure. The (D') and (%E) increased with growth of pH. The results indicate pH, (%E), and (D') decrease with increasing initial Cu(I) concentration (5×10^{-4} to 2.5×10^{-2}) M and they raise with increasing lauric acid concentration 0.2–0.5 M. The negative of $\Delta G°$ indicated that the Cu(I) extraction with lauric acid occurred spontaneously.

KEYWORDS

- **Copper sulfate pentahydrate**
- **Distribution coefficient**
- **Lauric acid**
- **Percentage extraction**
- **Thermodynamic temperature**

REFERENCES

1. Nermerow, N. L. *Industrial Waste Treatment: Contemporary Pravtice and Vision for the Future.* Elsevier Inc. United States of America (2007).

2. Trigg, G. L. and Edmund, H. *Immergut Encyclopedia of applied physics.* VCH Publishers (1992).

3. Csuros, M. and Csuros, C. *Environmental Sampling and Analysis for Metals.* CRC Press LLC, United States of America (2002).

4. Kurniawan, T. A., Chan, G. Y. S., Lo, W. H., and Babel, S. Physico-chemical treatment techniques for wastewater laden with heavy metals. *Chem. Eng. J*, **118**, 83–98 (2006).

5. Sud, D., Mahajan, G., and Kaur, M. P. Agricultural waste material as potential adsorbent for sequestering heavy metal ions from aqueous solutions – a review. *Bioresour. Technol.*, **99**, 6017–6027 (2008).

6. Cox, M. and Rydberg, J. In Introduction to solvent extraction. *Solvent Extraction Principles and Practice.* J. Rydberg and M. Cox, C. Musikas, and G. R. Choppin (Eds.). Marcel Dekker Inc., United States of America (2004).

7. Chang, S. H., Teng, T. T, and Ismail, N. Extraction of Cu(II) from aqueous solutions by vegetable oil-based organic solvents. *Journal of Hazardous Materials*, **181**, 868–872 (2010).

8. Chang, S. H., Teng, T. T., and Ismail, N. Efficiency, stoichiometry and structural studies of Cu(II) removal from aqueous solutions using di-2-ethylhexylphosphoric acid and tributylphosphate diluted in soybean oil. *Chemical Engineering Journal*, **166**, 249–255 (2011).

9. Alguacil, F. J., Cobo, A., and Alonso, M. Copper separation from nitrate/nitric acid media using Acorga M5640 extractant Part I: Solvent extraction study. *Chemical Engineering Journal*, **85**, 259–263 (2002).

10. Touati, S. and Meniai, A. H. Experimental Study of the Extraction of Cu(II) from Sulfuric Acid by Means of Sodium Diethyldithiocarbamate (SDDT). World Academy of Science. *Engineering and Technology*, **76**, 542–545 (2011).

11. Xie, F. and Dreisinger, D. Studies on solvent extraction of copper and cyanide from waste cyanide solution. *Journal of Hazardous Materials*, **169**, 333–338 (2009).

12. Kara, D. and Alkanb, M. Preconcentration and separation of Cu(II) with solvent extraction using N,N'-bis(2-hydroxy-5-bromo-benzyl) 1,2-diaminopropane. *Microchemical Journal*, **71**, 29–39 (2002).

13. Choppin, G. R. In Complexation of metal ions. *Solvent Extraction Principles and Practice.* J. Rydberg, M. Cox, C. Musikas, and G. R. Choppin (Eds.). Marcel Dekker Inc., United States of America (2004).

7 Modeling of Microwave Drying Process

CONTENTS

7.1 INTRODUCTION

Microwave drying has several advantages over conventional hot air drying, such as higher drying rate, minimal heating at locations with less water thus reducing overheating of locations where, heating is not required. However, for microwave drying to be more useful at the industrial level, it needs information on moisture diffusion models that could describe the process accurately. The diffusion coefficient of a food is material property and its value depends upon the conditions within the material. Effective moisture diffusivity describes all possible mechanisms of moisture movement within the foods, such as liquid diffusion, vapor diffusion, surface diffusion, capillary flow, and hydrodynamic flow (Kim and Bhowmik, 1995). Availability of such correlations and models, verified by experimental data, will enable engineers and operators to provide optimum solutions to aspects of drying operations such as energy use, operating conditions, process control, and without undertaking experimental trials on the system [9]. In particular, thin-layer Equations contribute to the understanding of the heat and mass transfer phenomena and computer simulations for designing new processes and improving existing commercial operations [12]. Thin-layer drying models can be categorized as theoretical, semi-theoretical, and empirical [5]. Models within the latter two categories consider only external resistance to moisture transfer [15] and neglect the effect of variation in sample temperature on the drying process [6]. The models are generally derived by simplifying general series solutions of Fick's second law and are only valid within the drying conditions for which they have been developed. However, they require short time, as compared with theoretical thin-layer equations, and do not require assumptions regarding sample geometry, mass diffusivity, and conductivity. Such models include the Page [1], Henderson and Pabis [2], two-term Sharaf-Eldeen, Blaisdell, and Hamdy [4], approximation of diffusion (Yaldiz and Ertekin [13]), and Midilli, Kucuk, and Yapar Equations [14].

Empirical models, which derive a direct relationship between moisture content and drying time, neglect the fundamentals of the drying process and have parameters with no physical meaning [16]. Among them, Wang and Singh [3] and Chavez-Mendez, Salgado-Cervantes, Garcia-Galindo, De La Cruz-Medina, and Garcia-Alvarado [7] have found application in literature. Kardum et al. [12] reported that the microwave drying kinetics of a pharmaceutical product was adequately described by the page mode with the latter providing a better correlation with the experimental data Abdelghani-Idrissi [11] approximated the transient behavior of normalized moisture during the microwave heating of cement powder by an exponential evolution with a time constant. In this chapter we used from experimental data of microwave drying of potato and determined the best mathematical thin-layer model to quantify the moisture removal behavior.

7.2 MATERIALS AND METHODS

7.2.1 Materials

Potatoes are also representative of a large range of high moisture foods and are also one of the main crops of major economic significance. Potato pieces bought from local market and then cut to six samples for drying tests. Dimensions of samples brought in Table 1.

TABLE 1 Summary of experiments.

Thickness (mm)	Diameter (mm)	Micro wave Power (w)	Pressure (kpa)
3	20	100	101
3	20	500	101
10	20	100	101
10	20	500	101
3	40	100	101
3	40	500	101

7.2.2 Equipment

The drying system used in this work is microwave oven (Butan, model no. MF 45) of variable power output settings, rated capacity of 900 W at 2.45 GHz, outside dimensions (W × D × H), 601 × 465 × 338 mm and cavity dimensions (W × D × H), 419 × 428 × 245 mm.

7.2.3   Experimental Method

The drying characteristics of potato during microwave processing for six samples were examined. Three samples with dimensions (thickness = 3 mm, diameter = 20 mm and t = 10, d = 20 and t = 3, d = 40) dried one time with power of microwave (100 W) and another (500 W). At 2 min intervals throughout the drying process (until material had attained at least 95% moisture loss) samples were removed and weighted before each experimental.

7.2.4   Data Analysis

The experimental moisture content data were non dimensionlized using the Equation:

$$MR = \frac{X - X_e}{X_0 - X_e} \tag{1}$$

where MR is the moisture ratio, X_0 is the initial moisture content, X_e (kg/kg) is the equilibrium moisture content, and X is the moisture content at time t.

$$X = \frac{M_t - M_d}{M_d} \tag{2}$$

where M_t = Sample mass at the time, M_d = mass of dried sample. For the analysis it was assumed that the equilibrium moisture content X_e was equal to zero. Selected thin-layer models, detailed in Table 2 were fitted to the drying curves (MR *versus* time), and the Equation parameters determined using non-linear least squares regression analysis. Two criteria were adopted to evaluate the goodness of fit of each model, the root mean square error (RMSE) and sum of squares due to error (SSE).

$$RMSE = [\frac{1}{N}\sum_{i=1}^{N}(MR_{exp,i} - MR_{pred,i})^2]^{0.5} \tag{3}$$

$$SSE = \sum_{i=1}^{N}(MR_{exp,i} - MR_{pred,i})^2 \tag{4}$$

where $MR_{exp,i}$ is the experimental moisture ratio, $MR_{pred,i}$ is the predicted moisture ratio, N is the number of experimental data points. The lower calculated values of RMSE and SSE, the better ability of the model to represent the experimental data. A value of RMSE and SSE closer to zero indicates a better fit. We take the constant coefficients to minimize the RMSE and SSE with Matlab software.

TABLE 2   Thin-layer models fitted to experimental data.

Model	Mathematical expression
Page	$MR = \exp(kt^n)$
Henderson and Pabis	$MR = a\,\exp(kt)$

TABLE 2   *(Continued)*

Model	Mathematical expression
Modified Henderson and Pabis	$MR = a \exp(kt) + b \exp(gt) + c \exp(ht)$
Logarithmic	$MR = a \exp(kt) + c$
Two-term	$MR = a \exp(k_1 t) + b \exp(k_2 t)$
Wang and Singh	$MR = 1 + at + bt^2$
Midilli	$MR = a \exp(kt^n) + bt$

Run the microwave was preheated at full power (100 W) for 5 min using a 500 ml water load.

7.3 DISCUSSION AND RESULTS

7.3.1 Drying Rate Constant and Effect of Variables

The experimental raw data represented in the change of weight with time form the basis of the drying kinetics calculations. The average moisture content of potato during drying was determined and expressed on a dry mass basis. The drying rate curves were then computed using the Lagrangian differentiation technique based on five consecutive data points [8]. For equally spaced time intervals, the five point equations are represented by Mapnson et al. [8].

$$\frac{dX_1}{dt} = \frac{[-25X_1 + 48X_2 - 36X_3 + 16X_4 - 3X_5]}{1200\Delta t} \tag{5}$$

$$\frac{dX_2}{dt} = \frac{[-3X_1 - 10X_2 + 18X_3 - 6X_4 - X_5]}{1200\Delta t} \tag{6}$$

$$\frac{dX_3}{dt} = \frac{[X_1 - 8X_2 + 8X_4 - X_5]}{1200\Delta t} \tag{7}$$

$$\frac{dX_4}{dt} = \frac{[-X_1 + 6X_2 - 18X_3 + 10X_4 + 3X_5]}{1200\Delta t} \tag{8}$$

$$\frac{dX_5}{dt} = \frac{[3X_1 - 16X_2 + 36X_3 - 48X_4 + 25X_5]}{1200\Delta t} \tag{9}$$

where X_{1-5} are moisture contents (kg^{-1}) at times t_{1-5} (m).

The smoothed drying rate curve is shown in Figure 1 and represents the drying rate for the three drying modes. It can be observed from Figure 1 that the drying rate curves are characterized by three distinct.

Drying periods is an initial heating up period representing a transition period corresponding to non-isothermal conditions, and two falling rate periods which exhibit an exponential variation in the residual moisture with time. Hence, each of the falling rate periods can be characterized by an exponential equation.

$$X = a\exp(-k_c t) \qquad\qquad (10)$$

where a is the pre-exponential factor (kg^{-1}) represents the initial moisture content and k_c is the exponentional variable, which gives an indication of the specific drying constant (min^{-1}). The drying rate constant, k_c can be calculated from Equation (10) for falling rate period at different drying conditions, the calculated results are presented in Table 3.

Figure 1 indicates an absence of a constant rate drying period which confirms that moisture removal driven internally by microwave energy absorption takes place exclusively in the falling rate period. After a short heating-up period the drying rate increases rapidly until it reaches a maximum before it decreases progressively. This is attributed to potatoes having high moisture content during the initial stages of drying. At this stage liquid vaporization may occur within the material causing a pressure drop to develop as this vapor expands and seeks to escape through the pores of the drying material since the pores are filled or partially filled with water. In the falling rate period, the moisture/vapor transfer outwards lags behind the evaporation rate at the surface. Microwave power absorption largely depends on the moisture content and as the product loses moisture the microwave absorption decreases progressively. This could be the reason for decreased drying rates during the later parts of drying. Figure 1 also indicates that an increase in the microwave power applied produced a steeper drying curve resulting in faster removal of moisture. Figure 2 shows the relationship between the drying rate constants at different microwave power levels. The effect of microwave power is reflected on the values of the drying constant, where a sharp increase in the drying constant was obtained at higher values of microwave power [10]. Also from Figure 2 realize that increase in thickness and diameter of potato samples, at constant power of microwave, the drying rate constant (k_c) increase.

7.4   MODEL APPLICATION

Thin-layer models have found wide application due to their ease of use and lack of required data, such as phenomenological and coupling coefficients, as in complex theoretical models. The experimental moisture content results were non-dimensionlized using Equation (1). The dimensionless data were the regressed against time, according to the form of the various thin-layer correlation Table 2, using the least squares curve fitting method. This defined the drying behavior in terms of the drying constants (k, k_1, k_2) and constant (a, b, c, g, h, n) as appropriate to the specific Equation. Table

4 and Table 5 details the parameter values for seven drying models with the corresponding (RMSE) and (SSE) values for representative drying techniques.

$RMSE_{ave}(page) = .02031$

$RMSE_{ave}(handerson \& pabis) = .03566$

$RMSE_{ave} (Modified\ Henderson\ and\ Pabis) = .02029$

$RMSE_{ave}\ logarithmic = .0186$

$RMSE_{ave}(Two-term) = 0.012559$

$RMSE_{ave}(Wang - \sin gh) = 0.0476$

$RMSE_{ave}(Midilli) = 0.0150$

$SSE_{ave}(page) = .0081345$

$SSE_{ave}(handerson \& pabis) = 0.024142$

$SSE_{ave} (Modified\ Henderson\ and\ Pubis) = .06476$

$SSE_{ave}{}^{\prime} (logarithmic) = 0.00545$

$SSE_{ave}(Two-term) = 0.00237$

$SSE_{ave} (Wang - Singh) = 0.04877$

$SSE_{ave}(Midilli) = 0.003547$

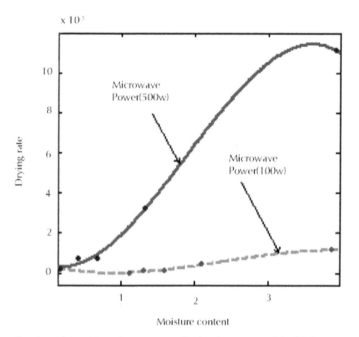

FIGURE 1   Drying rate curves for potato cylinder (power = 100, thickness = 3, and diameter = 20) under microwave drying.

TABLE 3 Drying rate constant for potato samples under microwave drying.

Microwave power (w)	Thickness(mm)	Diameter(mm)	k_c (1/min)	a (kg/kg)
100	10	20	0.01602	3.553
500	10	20	0.2225	3.599
100	3	20	0.01005	3.698
500	3	20	0.05751	3.986
100	3	40	0.0188	3.699
500	3	40	0.183	4.09

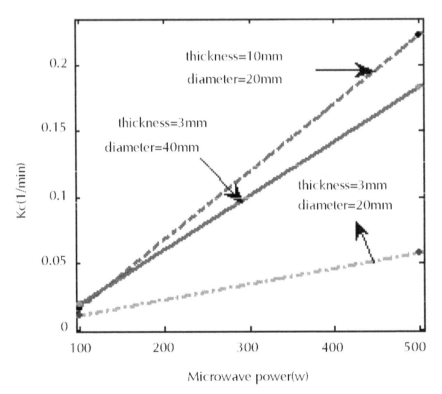

FIGURE 2 Effect of microwave power absorbed and sample diameter on the drying constant for potato.

TABLE 4   Estimated value of coefficients and statistical analysis for the thin-layer model (microwave power = 100 W.

Model	Constants	Thickness = 3(mm) Diameter = 20(mm)	Thickness = 10(mm) Diameter = 20(mm)	Thickness = 3(mm) Diameter = 40(mm)
Page	k	−.01388	−.0608	−.03174
	n	0.9273	.06778	.8636
	R^2	0.9995	.9745	.9902
	RMSE	0.004825	.03791	.02668
	SSE	0.000326	.02012	.009965
Henderson and Pabis	a	0.9859	.9157	.9786
	k	−.01005	−.01602	−.01879
	R^2	0.9984	.9151	.9809
	RMSE	0.008371	.06913	.03724
	SSE	0.000981	.0669	.01941
Modified Henderson and Pabis	g	0.05506	.5032	.5748
	b	0.9491	.479	.4773
	c	−0.004107	.07546	−.052
		−0.009403	−.006655	−.05328
	g	−0.1339	−.9088	−10.94
	h	−0.07544	−.06725	−.009982
	k	0.9995	.9964	.9989
	R^2	0.005564	.01779	.01037
	RMSE	0.0003096	.002847	.001074
	SSE			
Logarithmic	a	0.8787	.7071	.8393
	k	−0.01227	−.04408	−.02985
	c	0.1142	.3064	.1805
	R^2	0.9991	.9927	.997
	RMSE	0.006273	.02099	.01543
	SSE	0.0005115	.005727	.003096

TABLE 4 *(Continued)*

Model	Constants	Thickness = 3(mm) Diameter = 20(mm)	Thickness = 10(mm) Diameter = 20(mm)	Thickness = 3(mm) Diameter = 40(mm)
Two-term	a	0.06337	.4938	.4783
$\rightarrow$	k_1	−0.05237	−.06961	−.04717
	b	0.9347	.5373	.5512
	k_2	−0.009245	−.006133	−.009785
	R^2	0.9994	.9964	.9975
	$RMSE$	0.005236	.01542	.01457
	SSE	0.000329	.002854	.002548
$\leftarrow$	a	−0.01049	−.02026	−.01869
Wang and Singh	b	4.248×10^{-5}	.0001357	.0001103
	R^2	0.997	.9334	.9825
	$RMSE$	0.01132	.06124	.03563
	SSE	0.001793	.05251	.01777
	a	1.002	1.032	1.024
	k	−0.01461	−.05111	−.02936
Midilli	n	0.9047	.8184	.9458
	b	−0.0002144	.001916	.001164
	R^2	0.9995	.9912	.9963
	$RMSE$	0.005097	.02405	.01764
	SSE	0.0003118	.006942	.003734

TABLE 5 Estimated value of coefficients and statistical analysis for the thin-layer model (microwave power = 500 W).

Model	Constants	Thickness = 3(mm) Diameter = 20(mm)	Thickness = 10(mm) Diameter = 20(mm)	Thickness = 3(mm) Diameter = 40(mm)
	k	−.03223	−.3802	−.2598
	n	1.211	.7356	.8279
Page	R^2	.9978	.9982	.9871
	$RMSE$	.009668	.01119	.03161
	SSE	.001402	.002004	.01499

TABLE 5   *(Continued)*

Model	Constants	Thickness = 3(mm) Diameter = 20(mm)	Thickness = 10(mm) Diameter = 20(mm)	Thickness = 3(mm) Diameter = 40(mm)
Henderson and Pabis	a	1.044	.9133	.9645
	k	−.05754	−.2244	−.1829
	R^2	.994	.9773	.9758
	RMSE	.01592	.04005	.04328
	SSE	.003801	.02567	.02809
Modified Henderson and Pabis	g	8.008	3.144	12.45
	b	−7.061	.4012	−11.7
	c	.0813	−2.554	.1992
	g	.001265	−.7368	−.1234
	h	−.744	−.1735	−.3511
	k	−.003255	−.1691	−.125
	R^2	.9858	.9982	.9797
	RMSE	.02862	.01318	.04626
	SSE	.009012	.002085	.02353
Logarithmic	a	−59.55	.8916	.9182
	k	.0006566	−.2897	.1008
	c	60.55	.06441	−.2598
	R^2	.9911	.99	.9945
	RMSE	.02006	.02747	.0214
	SSE	.005635	.01132	.006411
Two-term →	a	1.085	.414	.7328
	k_1	−.06162	−.7293	−.3267
	b	−.08461	.5817	.295
	k_2	−.8359	−.1501	−.06892
	R^2	.9991	.9982	.995
	RMSE	.006778	.01214	.02121
	SSE	.0005973	.002064	.00585

TABLE 5 *(Continued)*

Model	Constants	Thickness = 3(mm) Diameter = 20(mm)	Thickness = 10(mm) Diameter = 20(mm)	Thickness = 3(mm) Diameter = 40(mm)
←	a	−.04384	−.1592	−.1496
Wang and Singh	b	.0003609	.006437	.006162
	R^2	.9945	.8605	.9487
	RMSE	.01533	.09919	.06303
	SSE	.003594	.1574	.05958
Midilli	a	1.002	1.002	1.017
	k	.09927	−.3847	−.2404
	n	.7863	.7232	.9624
	b	−.1253	−.0004	.005363
	R^2	.999	.9983	.9934
	RMSE	.007165	.01183	.02429
	SSE	.0006673	.001959	.00767

From this result the model with minimum of $RMSE_{ave}$ and SSE_{ave} is the best model for curve fitting of data sets. Two term model ($a\exp(k_1 t) + b\exp(k_2 t)$) is the best. Then Midilli model ($a\exp(kt^n) + bt$) Figure 3, and logarithmic model ($a\exp(kt) + c$) are better than others, respectively. Figure 4 represents experimental moisture ratio *versus* predicted one for all thin-layer models.

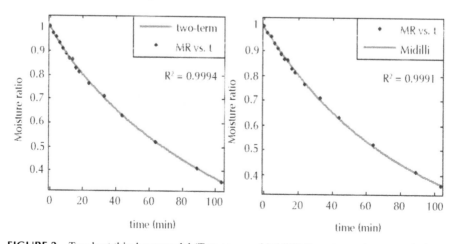

FIGURE 3 Two best thin-layer model (Two-term and Midilli) for potato microwave drying.

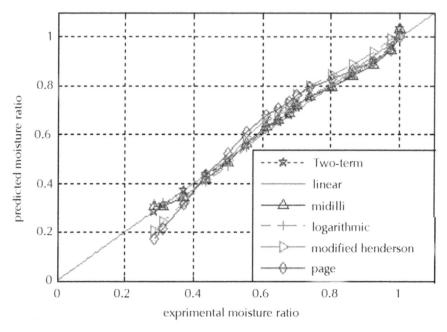

FIGURE 4 Experimental and predicted drying curves for potato sample (thickness = 10 mm, diameter = 20 mm) Dried under microwave (100 W).

7.5 CONCLUSION

(1) The drying rate constant (Kc), increase with increasing in microwave drying power.

(2) From two error criterion that discussed, we can conclude that the error of thin-layer modeling in porous media increase with increasing in thickness of samples.

(3) Of the seven thin-layer drying correlations considered, the two term and Midilli model provided the best representation of porous media microwave drying kinetics.

KEYWORDS

- **Empirical model**
- **Matlab software**
- **Microwave drying process**
- **Root mean square error**
- **Thin-layer models**

REFERENCES

1. Page, G. *Factors influencing the maximum rates of air drying shelled corn in thin layers*. M. Sc. Thesis, Purdue University, Indiana, USA (1949).
2. Henderson, S. M., and Pabis, S. Grain drying theory I: Temperature effect on drying coefficient. *Journal of Agriculture Research Engineering*, **6**, 169–174 (1961).
3. Wang, C. Y. and Singh, R. P. *A single layer drying equation for rough rice*. ASAE, Paper No 78-3001 St. Joseph, MI: ASAE (1978).
4. Sharaf-Eldeen, Y. I., Blaisdell, J. L., and Hamdy, M. Y. A model for air corn drying. *Transactions of American Society of Agricultural Engineers*, **23**, 1261–1265 (1980).
5. Parti, M. A theoretical model for thin-layer grain drying. *Drying Technology*, **8**, 101–122 (1990),
6. Parti, M. Selection of mathematical models for drying grain in thin-layers. *Journal of Agricultural Engineering Research*, **54**, 339–352 (1993).
7. Chavez-Mendez, C., Salgado-Cervantes, M. A., Garcia-Galindo, H. S., De La Cruz-Medina, J., and Garcia-Alvarado, M. A. Modeling of drying curves for some foodstuffs using kinetic Equation of high order. *Drying Technology*, **13**, 2113–2122 (1995).
8. Mapnson, L. W., Swain, T., and Tomalin, A. W. In uence of variety cultural conditions and temperature of storage on enzymatic browning of potato tubers. *Journal of Science of Food and Agriculture*, **14**, 673–684 (1963).
9. Dincer, I. Moisture loss from wood products during drying Part II: Surface moisture content distributions. *Energy Source*, **20**(1), 77–83 (1998).
10. Khraisheh, M. A. M., McMinn, W. A. M., and Magee, T. R. A. A multiple regressions approach to the combined microwave and air drying process. *Journal of Food Engineering*, (1999).
11. Abdelghani-Idrissi, M. A. Experimental investigations of occupied volume effect on the microwave heating and drying kinetics of cement powder in mono-mode cavity. *Applied Thermal Engineering*, **21**, 955–965 (2001),
12. Kardum, J. P., Sander, A., and Skansi, D. Comparison of convective, vacuum, and microwave drying chlorpropamide. *Drying Technology*, **19**(1), 167–183 (2001).
13. Yaldiz, O. and Ertekin, C. Thin layer solar drying of some vegetables. *Drying Technology*, **19**, 583–597 (2001).
14. Midilli, A., Kucuk, H., and Yapar, Z. A new model for single layer drying. *Drying Technology*, **20**, 1503–1513 (2002).
15. Panchariya, P. C., Popovic, D., and Sharma, A. L. Thin-layer modeling of black tea drying process. *Journal of Food Engineering*, **52**, 349–357 (2002).
16. Ozdemir, M. and Devres, Y. O. The thin layer drying characteristics of hazelnuts during roasting. *Journal of Food Engineering*, **42**, 225–233 (1999).

8 Modeling of Isothermal Vapor-liquid Phase Equilibria

CONTENTS

8.1 INTRODUCTION

The precise vapor-liquid equilibrium (VLE) data of binary mixtures like alcohol-alcohol are important to design many chemical processes and separation operations. The VLE investigations of binary systems have been the subject of much interest in recent years [1-9]. In the past, several authors have reported isobaric VLE data for the system of tert-butanol (TBA) + water [10-13]. Darwish and Al-Anber [14] have presented isobaric VLE data for the binary systems of TBA + water and TBA + isobutanol at 94.9 kPa.

We have recently reported the LLE data for ternary mixture of (water + TBA + 2-ethyl-1-hexanol) [15] and (water + n-butanol + 2-ethyl-1-hexanol) [16]. From the experimental results, 2-ethyl-1-hexanol was chosen as a very good solvent for recovering TBA and NBA from aqueous solution [17]. This solvent has been also used as an organic solvent to determination of LLE data for some ternary mixtures of (water + ethanol + 2-ethyl-1-hexanol) [18], (water + acetone + 2-ethyl-1-hexanol) [19], and (water + acetic acid + 2-ethyl-1-hexanol) [20].

In order to know the activity coefficients of the binary systems of TBA + 2-ethyl-1-hexanol and NBA + 2-ethyl-1-hexanol, the VLE data are also needed. These experimental data are determined using a method based on gas chromatographic (GC) technique described previously [17]. The experimental activity coefficients and VLE data can be correlated to several equilibrium methods such as universal quasi-chemical (UNIQUAC) model and NTRL models. Equilibrium models, such as the UNIQUAC [21] and the non-random two-liquid model (NRTL) [22] have been successfully applied for the correlation of several liquid-liquid and vapor-liquid systems. These models depend on optimized interaction parameters between each pair of components in the systems, which can be obtained by experiments.

In this research we have presented experimental activity coefficients and isobaric VLE data for the binary systems TBA + 2-ethyl-1-hexanol and NBA + 2-ethyl-1-hexanol at 101.3 kPa. The calculated activity coefficients have been correlated by fitting the binary interaction parameters of different models UNIQUAC and NRTL.

8.2 EXPERIMENTAL

8.2.1 Materials

The TBA, n-buatnol (NBA), 2-ethyl-1-hexanol were obtained from Merck a purity of 99.8% and were used without further purification.

8.2.2 Aparatus and Procedure

The quick method to measure VLE data has been a circulation one. These methods they are those usually employees for the direct determination VLE data because they are simple and precise enough the quantity of needed chemist is small. The VLE measurements were carried out using a Labdest unit built by Fischer, and equipped with a Cottrell pump was used in the equilibrium determinations.

The apparatus and procedure have been described completely by Artigas [13] and Walas [14]. The composition analysis of the vapor and liquid phase were determined using Konik GC equipped with a thermal conductivity detector (TCD) and Shimadzu C-R2AX integrator. The column injector and detector temperatures were 493.15, 404.15, and 433.15K respectively. A 2 m × 2 mm column was used to separate the components.

8.2.3 The UNIQUAC Model

At (VLE), the composition of the two-phases (vapor phase and liquid phase) can be determined from the following equations:

$$\sum y_i = \sum k_i x_i \tag{1}$$

$$\sum x_i = \sum y_i = 1 \tag{2}$$

$$\gamma_i = \frac{y_i P}{x_i p_i^0} \tag{3}$$

$$\frac{G^E}{RT} = \sum_{i=1}^{n} x_i \ln \gamma_i \tag{4}$$

Here k_i constant equilibrium, γ_i activity coefficient of component i, x_i, molar excess Gibbs energy G^E/RT and y_i the mole fraction of liquid phase and vapor phase are respectively the liquid and vapor phase compositions in equilibrium, p the total pressure, p_i^0 the pure component vapor pressure calculated by using the Antoine equation parameters between TBA and 2-ethyl-1-hexanol was correlated with activity coefficients. Equation (1), (2), and (3) are solved for the mole fraction (x) of component i in the two-phases. This method of calculation gives:

The UNIQUAC model is given by Abrams and Prausnitz [8] as:

$$\frac{g^E}{RT} = \sum_{i=1}^{c} x_i \ln(\frac{\Phi_i}{x_i}) + \frac{z}{2}\sum_{i=1}^{c} q_i x_i \ln(\frac{\theta_i}{\Phi_i}) - \sum_{i=1}^{c} q_i x_i \ln(\sum_{j=1}^{c}\theta_j\tau_{ji}) \tag{5}$$

or

$$\ln\gamma_i = \ln\gamma_i^c + \ln\gamma_i^R \tag{6}$$

where

$$\ln\gamma_i^c = \ln\left(\frac{\Phi_i}{x_i}\right) + \frac{z}{2}qi\,\ln\left(\frac{\theta_i}{\Phi_i}\right) + i_1 - \frac{\phi_i}{x_i}\sum_{j=1}^{c}x_j i_j \tag{7}$$

$$\ln\gamma_i^R = q_i\left[1 - \ln\left(\sum_{j=1}^{c}\theta_j\tau_{ji}\right) - \sum_{j=1}^{c}\frac{\theta_j\tau_{ij}}{\sum_{k=1}^{c}\theta_k\tau_{kj}}\right] \tag{8}$$

Here, γ_i^c is combinatorial part of the activity coefficient, and γ_i^R the residual part of the activity coefficient. The variable τ_{ij} adjustable parameter in the UNIQUAC equation and x_i the equilibrium mole fraction of component i. The parameter Φ_i (segment fraction) and θ_i (area fraction) are given by the following equation:

$$\Phi_i = \frac{x_i r_i}{\sum_{i=1}^{c} x_i r_i} \tag{9}$$

$$\theta_i = \frac{x_i r_i}{\sum\limits_{i=1}^{c} x_i q_i}$$
(10)

$$T_{ij} = \exp\left(-\frac{(u_{ij} - u_{jj})}{RT}\right) = \exp\left(-\frac{a_{ij}}{T}\right)$$
(11)

The parameter u_{ij} characterizes the interaction energy between compounds i and j and u_{ij} equals u_{ji}.

where a_{ij} is the UNIQUAC parameter with temperature independence and represents the energy interactions between an i-j pair of molecules.

$$l_i = \left(\frac{z}{2}\right)(r_i - q_i) - (r_i - 1)$$
(12)

where z = 10, is lattice coordination number, r_i the number of segments per molecule and q_i the relative surface area per molecule.

8.2.4   The Non Random Two Liquid (NRTL) Model

The NRTL equation is given by [15]. The parameters $(g_{ji}-g_{ii})$ and $(\alpha_{ji} = \alpha_j)$ can be obtained from binary data. For each possible binary pair in a mixture, three parameters are need. The NRTL equation may be used to represent vapor-liquid and liquid-liquid equilibria [16].

$$\frac{g^E}{RT} = \sum_{i=1}^{n} x_i \left[\sum_{j=1}^{n} x_{ji} T_{ji}\right]$$
(13)

where

$$x_{ji} = \frac{x_j \exp\left(-\alpha_{ji} T_{ji}\right)}{\sum\limits_{k=1}^{n} x_k \exp\left(-\alpha_{ki} T_{ki}\right)}$$
(14)

or

$$\ln \gamma_i = \frac{\sum\limits_{j=1}^{n}\left(T_{ji} G_{ji} x_i\right)}{\sum\limits_{k=1}^{n}\left(G_{ki} x_k\right)} + \sum\limits_{j=1}^{n} \frac{x_j G_{ij}}{\sum\limits_{k=1}^{n} G_{kj} x_k}\left[T_{ij} - \frac{\sum\limits_{k=1}^{n} x_k T_{kj} G_{kj}}{\sum\limits_{k=1}^{n} G_{kj} x_k}\right]$$
(15)

$$G_{ij} = \exp\left(-\alpha_{ij} T_{ij}\right)$$
(16)

where

$$T_{ij} = \frac{\Delta g_{ij}}{RT}$$
(17)

8.3 DISCUSSION AND RESULTS

Experimental vapor-equilibria of TBA and NBA with 2-ethyl-1-hexanol were deter-mined and they are shown in Table 1 and Table 2. Isothermal vapor-liquid measure-ments were determined at 404.15 and 433.15 K for TBA+2-ethyl-1-hexanol (2EH) and NBA + 2EH. Table 3 Antoine coefficients were obtained by applying the equation:

$$\ln p_i^0 = A - \frac{B}{T-C} \qquad (18)$$

TABLE 1   Experimental VLE data for the system TBA 2-ethyl-1-hexanol (2EH).

T(K)	P(kP)	X1	Y1
404.15	34.1420	0.0648	0.6754
	77.2000	0.1477	0.8382
	133.8400	0.2542	0.9111
	211.3570	0.4094	0.9532
	261.9510	0.5098	0.9680
	297.9007	0.5810	0.9756
	311.7387	0.6093	0.9782
	367.8096	0.7222	0.9866
	425.5519	0.8380	0.9931
	478.7358	0.9937	0.9979
433.5			
	71.0210	0.0667	0.6430
	161.8159	0.1551	0.79344
	279.780	0.2703	0.8842
	442.1638	0.2998	0.9380
	548.2280	0.5353	0.9575
	622.9930	0.5967	0.9676
	652.6838	0.6393	0.9709
	770.3598	0.7569	0.9821
	891.5400	0.8764	0.9908
	960.7760	0.9479	0.9950

TABLE 2   Experimental VLE data for the system NBA 2-ethyl-1-hexanol (2EH).

T(K)	P(kP)	X1	Y1
404.2	25.8768	0.0667	0.3602
	58.6807	0.1514	0.5839
	101.6052	0.2630	0.7365
	160.1633	0.4164	0.8474
	198.2441	0.5170	0.8923
	225.0019	0.5880	0.9166
	235.6078	0.6162	0.9250
	277.5086	0.7297	0.9532
	320.4147	0.8380	0.9931
	478.7358	0.9937	0.9979
433.2			
	25.8768	0.0734	0.3962
	58.6707	0.1665	0.6423
	101.6051	0.2823	0.8102
	160.1639	0.4580	0.9321
	168.2441	0.5687	0.9369
	225.0019	0.6468	0.9629
	235.6078	0.6768	0.9712
	277.5140	0.8027	0.9818
	340.4172	0.9267	0.9952
	344.7790	0.9828	0.9964

Vapor pressures, p_i^0 (mmHg), were fitted with the Antoine equation the parameters A, B, and C of pure component are reported in Table 3 when the sample of vapor injected in the GC separating in all their components. The area of each resulting pick is proportional to the mole fraction of these components. The Figure 1 shows activity coefficients γ_i that were calculated from the equation:

$$Ai = Kini \tag{19}$$

TABLE 3   Antoine coefficients by Equation (19).

Component	A	B	C
n-Butanol	17.2160	3137.02	−94.43
Tert-Butanol	16.8548	2658.29	−95.50
2ethyl-1-hexanol	15.3614	2773.46	−140.00

where A_i is the area of the pick of component i and n_i they are the moles of the component i K_i is a constant of proportionally, different for each component. This is certain if the detector has a lineal answer and the conditions of work of the column and detecting they stay constant.

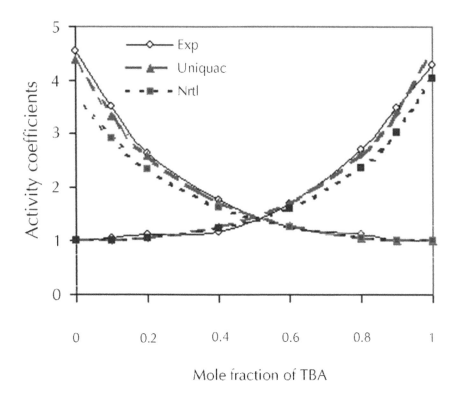

FIGURE 1   Activity coefficients in the system TBA – 2-ethyl-1-hexanol.

For each species i, the one numbers of moles of the phase vapor obtained in this analysis, it can be related with the fugacity of component i in the phase vapor, f_i^v, applying the following truncated equation of initial state:

$$f_i^v = \frac{n_i RT}{V} \exp\left[\frac{2}{v}\sum_j^m y_i B_{ij}\right] \tag{20}$$

where v is volume of the sample v it is the volume i specify, n_t total moles and B_{ij} is the 2° coefficient of the virial for even i and j.

We are concerned with a liquid mixture which, at temperature T and pressure P, is in equilibrium with a vapor mixture at the same temperature and pressure.

For every component I in the mixture, the condition of thermodynamic equilibrium is given by:

$$f_i^l = f_i^v \tag{21}$$

where f_i^l and f_i^v are fugcities liquid and vapor. The Equation (23) is completed for each present component in the two-phases in equilibruim. If the phase liquid is a pure component, the fugacity of pure liquid f_i^0 by:

$$f_i^0 = f_i^v \tag{22}$$

Activity coefficient γ_i is related to x_i and to fugacity f_i^0 (22)

$$\gamma_i = \frac{a_i}{x_i} = \frac{f_i^l}{x_i f_i^0} \tag{23}$$

where a_i is the activity of component i. The standard state fugacity f_i^0 is the fugacity of component i at the temperature of the system, that is the mixture and at some arbitrarily chosen pressure and composition. The choice of standard state pressure and composition is dictated only by convenience, but it is important to bear in mind that the numerical values of γ_i and a_i have no meaning unless f_i^0 is clearly specified.

Relationship of fugacities can be related with the pick area obtained starting from the analysis of the phase vapor. Making the substitution is appropriate: the activity coefficients can be expressed in function of the areas measured in the chromatogram in the following form:

$$\gamma_i = \frac{A_i}{x_i A_i'}\left\{\frac{\exp\left[\left(\frac{2}{V}\right)\sum_j^m n_j B_{ij}\right]}{\exp\left[\left(\frac{2}{V}\right)\sum_j^{m'} n_j B_{ij}\right]}\right\} \tag{24}$$

The exponential terms of the Equation (24) not only spread to be compensated mutually, but rather also each one spreads to the unit, the following expression is obtained for the coefficients of activities that were calculated from the equation, in Figure 1 show the activity coefficients:

$$\gamma_i = \frac{A_i}{x_i A_i'} \tag{25}$$

The parameters of the UNIQUAC and NTRL equations were optimized by minimizing the following objective function [12-18]:

$$OF = \sum_{j=1}^{np} \sum_{i=1}^{nc} \left[\frac{\gamma_{ij}^{exp} - \gamma_{ij}^{cal}}{\gamma_{ij}^{exp}} \right]^2 \tag{26}$$

where γ_{ij} are the corresponding activity coefficients and n_p is the number of experimental data and n_c is number components. In Table 4 and Table 5 shows the calculated value of the UNIQUAC and NRTL binary interaction parameters for the mixture TBA-2EH and NBA-2EH using universal values for the UNIQUAC and NRTL parameters, the mixture non-randomness parameter α_{12} in the NRTL equation was fixed at 0.3. The values of r and q used in the UNIQUAC equation are presented in Table 8.

TABLE 4   UNIQUAC and NRTL parameters the systems TBA + 2EH and NBA + 2EH.

I	J	UNIQUAC		NRTL	
		a_{12}	a_{21}	Δg_{12}	Δg_{21}
TBA	2-ethyl-1-hexanol	−55.4544	215.7780	511.769	480.0741
NBA	2-ethyl-1-hexanol	−78.3030	247.1739	384.4047	578.2420

TABLE 5   The UNIQUAC structural parameters.

Components	R	Q
TBA	3.45	3.05
NBA	3.45	3.05
2-ethyl-1-hexanol	6.15	5.02

The goodness of fit between the observed and calculated mole fractions was calculated in terms of RMSD. The RMSD values were calculate according to the equation of percentage RMSD (RMSD%):

$$RMSD\% = 100 \sqrt{\frac{1}{N} \sum_{i=1}^{N} \left(\frac{\gamma_i^{exp} - \gamma_i^{cal}}{\gamma_i^{exp}} \right)^2} \qquad (27)$$

where N is the number of data points, γ_i^{exp} indicates the experimental activity coefficient, γ_i^{cal} the calculated activity coefficient.

The predictions from UNIQUAC model, average (RMSD%) between the observed and calculated mol percents with a reasonable error was 4.71% for TBA + 2EH and 7.78% for NBA – 2EH and NRTL equation predicted its average RMSD value 6.43% for system TBA + 2EH and 9.6% for system NBA + 2EH. the values of RMSD UNIQUAC are small that those of NRTL, for that reason it is favorable to use of prediction VLE with model UNIQUAC that NRTL (see Table 8).

The experimental and predicted activity coefficient has been compiled in Table 6 and Table 7. The experimental and predicted VLE data for the binary system represented in Figure 1 and Figure 2. The solid line in these diagrams from experimental data in the predicted lines dashed from the UNIQUAC and NRTL equations. The isothermal vapor-liquid measurements were at 404.15 and 433.15K for TBA + 2-ethylhexanol and NBA + 2-ethy-1-hexanol Figure 5 and Figure 6.

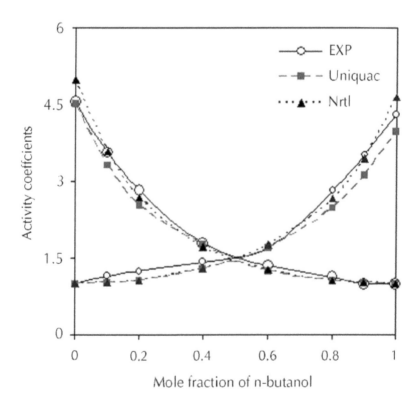

FIGURE 2   Activity coefficients in the system NBA – 2-ethyl-1-hexanol.

TABLE 6   Experimental and predicted activity coefficient for the binary system (TBA + 2EH).

Activity coefficient TBA			Activity coefficient 2-ethyl-1-hexanol		
Experimental	UNIQUAC	NRTL	Experimental	UNIQUAC	NRTL
4.7086	4.3899	4.2981	1	1	1
2.5696	3.3237	2.3596	1.1710	1.025	1.107
2.2514	2.5914	2.0012	1.1901	1.05	1.112
1.77699	1.7175	1.9945	1.3111	1.2538	1.229
1.29256	1.2759	1.1980	2.0621	1.8976	1.83
1.05999	1.0639	1.1901	2.7321	2.642	2.5
1.0072	1.0158	1.0789	3.2811	3.1053	3.297
1	1.0001	1.0003	4.6721	4.52	4.46
RMSD%	3.14	4.81		6.28	6.43

TABLE 7   Experimental and predicted activity coefficient for the binary system (NBA + 2EH).

Activity coefficient NBA			Activity coefficient 2-ethyl-1-hexanol		
Experimental	UNIQUAC	NRTL	Experimental	UNIQUAC	NRTL
4.5500	4.5182	4.991	1.0000	1	1
3.5700	3.3107	3.5758	1.1500	1.0164	1.0175
2.8200	2.5293	2.6902	1.2500	1.0657	1.0698
1.7720	1.7547	1.7248	1.4100	1.2772	1.2931
1.3500	1.2416	1.273	1.6900	1.7005	1.7511
1.1220	1.0537	1.063	2.8200	2.49	2.6672
1.0000	1.0129	1.0156	3.5200	3.1131	3.4555
1.0000	1	1	4.3000	3.9704	4.6419
RMSD%	5.74	10.28		9.82	8.92

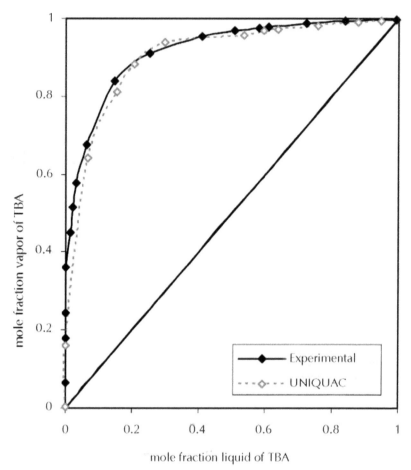

FIGURE 3   Calculated and observed vapor-liquid equilibria for TBA (1) – 2-ethyl-1-hexanol (2).

Figure 5 shows that the solubility of water in 2-ethyl-1-hexanol increases with amounts of 1-butanol added to water + 2-ethyl-hexanol mixture. The temperature in the range of the study has only a small effect on the solubility of water in 2-ethyl-1-hexanol.

8.3.1   Simulation and Separation Process

A commercial simulation program, ChemCad III, was used for simulation of the fractional distillation column. The flow diagram for NBA purification process is shown in Figure 6. The product (NBA) from the catalytic hydration of isobutene to TBA on Amberlyst-15 in a cocurrent down flow trickle-bed reactor [24] is sent to an extraction column prior to stripper. The NBA concentrations used in our design are 20 and 80 wt% water [24]. By removing water from the product flow, the NBA concentration in the top of the distillation column is 99.6 wt%. In this purification process, the production of NBA is estimated to be about 21.5 tons per day [25]. The stripper column was optimized at 27 plates with feed entering at plate 17. The total height of this column is 5 m and stripper diameter is 2 m. The values for optimization were estimated from the method of McCabe [26], and Coulson and Richardson [27] and also described by Dadgar and Foutch [25]. Details of process for the NBA purification set together with operation conditions (T, P) are presented in Table 9. The operation conditions were selected using the VLE data. The stream identification numbers in Table 9 and those of that in Figure 6 are the same. It should be noted that, in this table concentration is in terms of weight fraction. The simulation indicated recovery of the NBA by extraction to be facile. The enthalpy balance can be calculated using material balance, temperature, and thermodynamics data [27, 28].

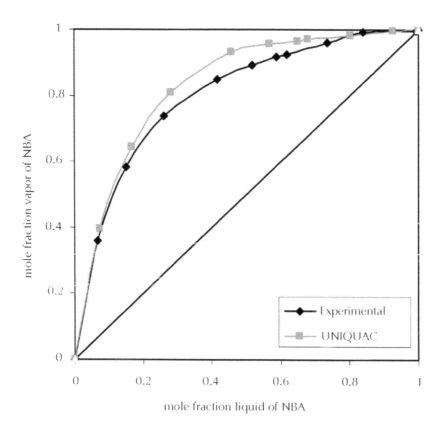

FIGURE 4   Calculated and observed vapor-liquid equilibria for NBA (1) – 2ethyl-1-hexanol (2).

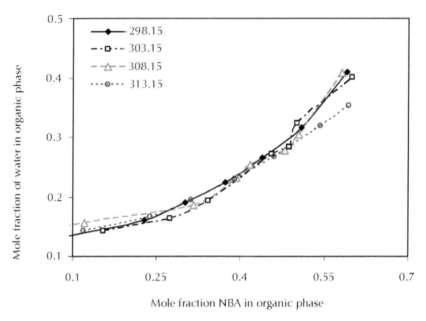

FIGURE 5 Effects of 1-butanol addition on solubility of water in 2-ethyl-1-hexanol at different temperatures.

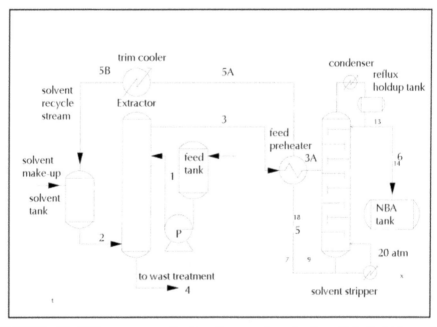

FIGURE 6 The NBA process separation flow diagram using extraction with 2-ethyl-1-hexanol.

TABLE 8 RMSD% values for the UNIQUAC and NRTL.

	NBA	TBA
UNIQUAC	5.74	3.14
	9.82	6.28
Average	7.78	4.71
NRTL	10.28	4.81
	8.92	8.05
Average	9.6	6.43

TABLE 9 Process for the NBA purification set together with operation conditions.

Currents of the process

	1	2	3	4	3A	5	5A	5B	6
NBA	0.180	------	0.151	0.0001	0.151	0.0005	0.0005	0.0005	0.9995
water	0.820	------	0.001	0.9999	0.001	0.0000	0.0000	0.0000	0.000
2EH	-----	1.0	0.848	0.0000	0.848	0.9995	0.9995	0.9995	0.0005
Flow (Kg/h)	5000	8000	5941.6	7058.4	5941.6	5046.5	5046.5	5046.5	895.1
Temperture (^{0}C)	25.	25.0	25.0	25.0	180	185	50.0	25.0	25
Pressure (atm)	1	1.0	2.7	1.0	2.0	1.0	2.0	1.0	1.0

8.4 CONCLUSION

The VLE data have been reported for TBA + 2EH and NBA + 2EH a two different temperatures at pressure 101.3 kPa.

Both binary system derivate very slight from ideal behavior. Experimental values, the Antoine constants were determined to enable the calculation of the saturated vapor pressures at a given temperature. The activity coefficients obtained by method of gas chromatografic are been that are thermodynamically consistent been correlate with UNIQUAC and NRTL equations.

The optimize interaction parameter were used to calculates the activity coefficients at same conditions. This gave an average RMSD values were 5.75, 4.28, 8.44, and 12.68% for the system of TBA + 2EH and NBA + 2EH for UNIQUAC and NRTL models respectively.

KEYWORDS

- **Antoine coefficients**
- **Gas chromatographic**
- **Non-random two-liquid model**
- **Universal quasi-chemical model**
- **Vapor-liquid equilibrium**

REFERENCES

1. Ghanadzadeh, H. and Ghanadzadeh, A. *Fluid Phase Equilibria*, in press (2002).
2. Ghannadzadeh, H. *Election of solvent selective for the extraction in phase liquido alcohols C4 (ABE) starting from biomass.* Ph.D. Thesis, Polytechnic Universitat of catalunya Barcelona, Spain (1993).
3. Quitzsch, K., Koop, R., Renker, W., and Geiseler, G. *Z. Phys. Chem.*, 237–265 (1968).
4. Suska, J. E., Holub, R., Vonka, P., and Pick, J. *Coll. Czech. Chem. Commun.*, **35**, 385 (1970).
5. Wichterle, J. linek and Hala, E. *Vapor-liquid equilibrium data Bibliography*. Elsvier, New York, (1982).
6. Knapp, H., Doring, R., Oellrich, L., Plocker, U., and Prausnitz, J. M. *Vapor_Liquid Equilibria for Mixtures of low Boiling Substances, Vol.VI*. Chemistry data series, DECHEMA, Frankfurt am Main (1981).
7. Gmehling, J. and Onken, U. *Vapor-liquid equilibrium data Collection, Vol.1*. Chemistry data series, DECHEMA, Frankfurt am Main (1977).
8. Hirata, M., Ohe, S., and Nagakama, K. *Computer-Aided data book of vapor–liquid equilibria*. Elsevier, New York (1975).
9. Kojima, K. and Tochigi, K. *Prediction of vapor-liquid equilibria by the ASOG Method*. Elsevier, Tokyo (1979).
10. Wilson, G. M. *J. Am Chem.*, **72**, 723 (1910).
11. Renon, H. and Prausnitz, J. M. *AIChE J*, **14**, 135 (1968).
12. Abrams, D. S. and Prausnitz, J. M. *AICHE J*, **21**, 116 (1975).
13. Artigas, H., Lafuente, C. Lopez, M. C., Royo, F. M. and Urieta, J. S. *Fluid Phase Equilib.*, **134**, 163 (1997).
14. Walas, S. M. *Phase Equilibria in chemical Engineering*, Butterworths, London (1985).
15. Henly, E. J. and Seader, J. D. *Equilibrium stage separations in chemical Engineering*. John Wiley & Sons, New York (1981).
16. Fredenslund, A., Gmehling, J., and Rasmussen, P. *Vapor-liquid equilibria*. Using UNIFAC, Elsevier Sci. Pub., New York, p. 285 (1977).
17. Reid, R. C., Prausnitz, J. M., and Poling, B. E. *The properties of Gases and Liquids*. McGraw-Hill, New York (1987).
18. Rodriguez, A., Canosa, J., Domenguez, A., and Tojo, J. *Fluid Phase Equilib.* **198**, 95 (2002).

9 Some Aspects of Liquid Phase Equilibrium at Different Temperatures

CONTENTS

9.1   INTRODUCTION

Liquid-liquid equilibrium data of ternary systems are required for the design of liquid extraction processes. Also, there is a constant need for phase equilibrium data of these systems for simulation and optimize of separation equipment, valuable information about the molecular interactions, macroscopic behavior of fluid mixtures, and can be used to test and improve thermodynamic models for calculating and predicting fluid phase equilibria. A large amount of investigation has been carried out in recent years on the LLE measurements of ternary systems, in order to understand and provide further information about the phase behavior of such systems [1-6].

In this work, the LLE data for the ternary system of (water + 1-hexanol + TBA) at temperatures from (298.15 to 305.15K) are presented. Here, TBA is used as a solvent in the separation of 1-hexanol from water. Complete phase diagrams are obtained by solubility and tie-line data simultaneously for each temperature. Selectivity values (S) are also determined from the tie-line data to establish the feasibility of the use of these liquid for the separation of (water + 1-hexanol) binary mixture. The experimental LLE data are correlated using the universal quasi-chemical (UNIQUAC).

9.2   EXPERIMENTAL

9.2.1   Materials

All chemicals used in this work, 1-hexanol and TBA, obtained by Merck with puri-
ties > 99% and were used without further purification. The purity of these materials
checked by gas chromatography. Deionized water was further distilled before use.

9.2.2   Apparatus and Procedure

The binoda (solubility) curves were determined by the cloud point method in an equilibrium glass cell (similar to that of Peschke and Sendler), Figure 1, connected to
a thermostat was made to measure the LLE data. The temperature of the cell was
controlled by a water jacket and maintained with an accuracy of within ±0.1K. The
mixture was vigorously agitated by a magnetic stirrer for 4 hr. The prepared mixtures
were then left to settle for 4 hr for phase separation. The samples of organic-rich phase
were taken by a syringe (1 µl) from the upper layer and that of water-rich phase from
a sampling tap at the bottom of the cell.

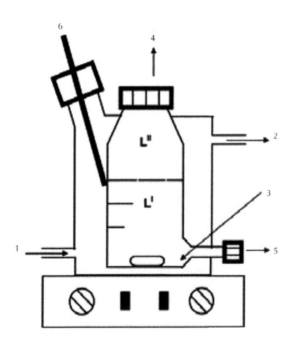

FIGURE 1   Glass cell.

The tie-line data were obtained by preparing ternary mixtures of known overall
compositions lying within the two-phase region, and after being allowed to reach
equilibrium, samples were carefully taken from each phase and analyzed. Both the
phases were analyzed using a Varian CP-3,800 gas chromatography (GC) equipped
with a thermal conductivity detector (TCD) and Star integrator. A 4 m × 4 mm column

packed with CHROMOSORB T 40–60 Mesh was used to separate the components. The injection and the detector temperatures were 250K. The carrier gas (helium) flow rate was maintained at 40 ml/min.

9.3 THE UNIQUAC MODEL

The experimental LLE data of a ternary system can be correlated using the UNIQUAC [7-9]. The mole fractions x_i^E, x_i^R of LLE phases (extracted phase and raffinate phase) can be determined using the following equation:

$$(\gamma_i x_i)^E = (\gamma_i x_i)^R \tag{1}$$

$$\sum x_i^\epsilon = \sum x_i^R = 1 \tag{2}$$

Here γ_i^E and γ_i^R are the corresponding activity coefficients of component i in extracted phase and raffinate phase. The UNIQUAC equation for the liquid-phase activity coefficient is represented as follows:

$$\ln \gamma_i = \ln \gamma_i(\text{combinatorial}) + \ln \gamma_i(\text{residual}) \tag{3}$$

The combinatorial and residual parts of the activity coefficient are due to difference in shape and energy of the molecules, respectively. The combinational and residual parts of the coefficient can be written as follows:

$$\ln \gamma_i^C = \ln\left(\frac{\Phi_i}{x_i}\right) + \frac{z}{2} q_i \ln\left(\frac{\theta_i}{\Phi_i}\right) + \iota_i - \frac{\phi_i}{x_i}\sum_{j=1}^{C} x_j \iota_j \tag{4}$$

$$\ln \gamma_i^R = q_i \left[1 - \ln\left(\sum_{j=1}^{c}\theta_j\tau_{ji}\right) - \sum_{j=1}^{c}\frac{\theta_j\tau_{ij}}{\sum_{k=1}^{c}\theta_k\tau_{kj}} \right] \tag{5}$$

Here τ_{ij} is the adjustable parameter in the UNIQUAC equation. The parameter Φ_i (segment fraction), θ_i (area fraction), and τ_{ij} are given by the following equations:

$$\Phi_i = \frac{x_i r_i}{\sum_{i=1}^{c} x_i r_i} \quad \text{and} \quad \theta_i = \frac{x_i r_i}{\sum_{i=1}^{c} x_i q_i} \tag{6}$$

$$T_{ij} = \exp\left(-\frac{\text{Äu}_{ij}}{RT}\right) = \exp\left(-\frac{a_{ij}}{T}\right) \tag{7}$$

The experimental results were compared with those correlated using the UNIQUAC model and the values for the interaction parameters were obtained for this model. The UNIQUAC parameters were estimated by Aspen. The UNIQUAC model has been successfully applied for the correlation of several LLE systems. This model depends on optimized interaction parameters between each pair of components in the system, which can be obtained by experiments. As the optimized interaction parameters can be also correlated to temperature, the interaction parameters in the UNIQUAC model were also investigated.

9.4   RESULTS

The LLE measurements for the ternary system were made at atmospheric pressure in the temperature range at (298.2, 303.2, and 305.2 2) K. The experimental and correlated LLE data of water, 1-hexanol and TBA at each temperature are obtained. Experimental tie-line data for (water + 1-heaxol + TBA) at each temperature were reported in Table 1.

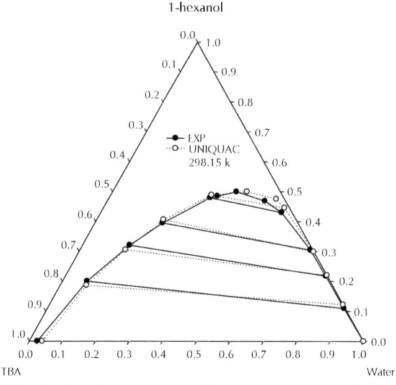

FIGURE 2   Experimental (—) and correlated UNIQUAC (---) LLE data at 298.15K.

TABLE 1 Experimental tie-line data for (water + 1-hexanol+ TBA) at each temperature.

Water-rich (mole fraction) phase			Solvent-rich (mole fraction) phase		
Water	1-hexanol	TBA	Water	1-hexanol	TBA
T=298.2K					
0.9999	0.0000	0.0001	0.0245	0.0000	0.9755
0.8867	0.1098	0.0035	0.0750	0.2001	0.7249
0.7774	0.2178	0.0048	0.1407	0.3210	0.5385
0.6880	0.3048	0.0072	0.2026	0.3950	0.4014
0.5389	0.4300	0.0311	0.3005	0.4798	0.2197
0.4691	0.4691	0.0618	0.3194	0.4855	0.1951
T=303.2K					
0.9860	0.0000	0.0140	0.0204	0.0000	0.9796
0.9380	0.0611	0.0009	0.0973	0.1639	0.7388
0.7573	0.2409	0.0018	0.2104	0.3354	0.4542
0.6942	0.3009	0.0049	0.2541	0.3766	0.3693
0.6292	0.3594	0.0114	0.3187	0.4130	0.2683
0.5414	0.4169	0.0417	0.3590	0.4273	0.2137
T=305.2K					
0.9948	0.0000	0.0052	0.0303	0.0000	0.9697
0.8816	0.1125	0.0059	0.0934	0.1744	0.7322
0.7716	0.2213	0.0071	0.1427	0.2768	0.5805
0.6897	0.3019	0.0084	0.2090	0.3540	0.0437
0.6145	0.3736	0.0119	0.2936	0.4106	0.2958
0.4959	0.4474	0.0567	0.3659	0.4464	0.1877

The experimental and correlated tie lines for this system at 298.2K were plotted in Figure 2. From the LLE phase diagram, (TBA+ water) is the only pair that is partially miscible and two liquid pairs (water + 1-hexanol) and (1-hexanol + TBA) are completely miscible. As it can be seen from Figure 2, the phase diagram shows plait point. At this point, only one liquid phase exists and the compositions of the two-phases are equal.

To show the selectivity and strength of the solvent in extracting the acid, distribution coefficients (D_i) for the 1-hexanol ($i = 2$) and water ($i = 1$) and the separation factor (S) is determined as follows:

$$D_i = \frac{\text{Weight fraction in solvent phase } (W_{i2})}{\text{Weight fraction in solvent phase } (W_{i1})} \tag{9}$$

$$S = \frac{(D_2)}{(D_1)} \tag{10}$$

The distribution coefficients and separation factor for each temperature are given in Table 2. The extraction power of the solvent at each temperature, plot of S $vs.$ X_{21}, is given in Figure 3.

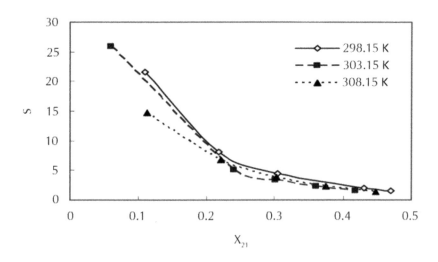

FIGURE 3   Separation factor (S) of 1-hexanol as a function of the mole fraction (X_{21}) of 1-hexanol in aqueous phase.

The consistency of experimentally measured tie-line data can be determined using the Othmer and Tobias correlation [10] for the ternary system at each temperature:

TABLE 2 Distribution coefficients (D_i) of water (1) and 1-hexanol (2) and separation factors at 298.15, 303.15 and 305.15K.

| 298.15 K | | | 303.15 K | | | 305.15 K | | |
D_2	D_1	S	D_2	D_1	S	D_2	D_1	S
1.8224	0.0846	21.5457	2.6825	0.1037	25.8600	1.5502	0.1059	14.6325
1.4738	0.1810	8.1432	1.3923	0.2778	5.0113	1.2508	0.1849	6.7632
1.2959	0.2945	4.4008	1.2516	0.3660	3.4193	1.1726	0.3030	3.8695
1.1158	0.5576	2.001	1.1491	0.5065	2.2687	1.0990	0.4778	2.3003
1.0350	0.6809	1.5200	1.0249	0.6631	1.5457	0.9978	0.7379	1.3523

$$\ln\left(\frac{1-X_{33}}{X_{33}}\right) = A + B \ln\left(\frac{1-X_{11}}{X_{11}}\right) \tag{11}$$

where A and B are constant, X_{33} is mole fraction of TBA in the extracted phase (organic-rich phase) and X_{11} is the mole fraction of water in the raffinate phase (aqueous-rich phase). Othmer–Tobias plots were presented in Figure 4 for the system at several temperatures and the correlation parameters ($R^2 \approx 1$) ≈ 1) are listed in Table 3. The linearity of the plots indicates the degree of reliability of the related data.

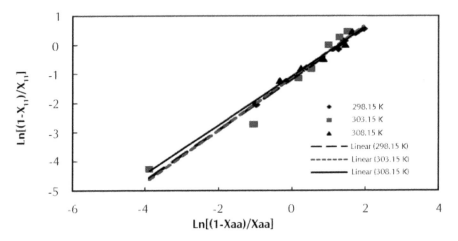

FIGURE 4   Othmer–Tobias of the (water + 1-hexanol +TBA) ternary system at different temperatures.

TABLE 3   Othmer-Tobias equation constants for (water + 1-hexanol + TBA) ternary system.

	Othmer–Tobias correlation		
T (K)	**A**	**B**	**R²**
298.2	0.8720	−1.1693	0.9962
303.2	0.9005	−1.1252	0.9579
305.2	0.8396	1.0761	0.9786

The root-mean-square deviation (RMSD) can be taken as a measure of precision of the correlations. The RMSD was calculated from the difference between the experimental and calculated mole fractions according to the following equation:

$$RMSD\% = 100\sqrt{\dfrac{\sum_{k=1}^{n}\sum_{j=1}^{2}\sum_{i=1}^{3}\left(\hat{X}_{ijk}-X_{ijk}\right)^2}{6n}} \tag{12}$$

where n is the number of tie-lines, x and $\hat{x}$ indicate the experimental and calculated mole fraction, respectively. The subscript i = indexes components, j = indexes phases and k = 1, 2,. . . ,n (tie-lines). The UNIQUAC model was used to correlate the experimental data at each temperature (298.15, 303.15, and 305.15K) with RMSD% values of 1.42, 1.97, and 1.33%, respectively.

9.5 CONCLUSION

The LLE data of the ternary system composed of water + 1-hexanol + TBA were measured at different temperatures of (298.2, 303.2 and 305.2) K. The UNIQUAC model was used to correlate the experimental LLE data. The optimum UNIQUAC interaction parameters between water, 1-hexanol, and TBA were determined using the experimental liquid-liquid data. The average RMSD value between the observed and calculated mole fractions with a reasonable error was 1.57% for the UNIQUAC model. The solubility of water in TBA increases with amounts of 1-hexanol added to water + TBA mixture.

KEYWORDS

- Gas chromatography
- Liquid phase equilibrium
- Othmer–Tobias equation
- Root-mean-square deviation
- Universal quasi chemical

REFERENCES

1. Garcia–Flores, B. E., Galicia-Aguilar, G., Eustaquio-Rincon, R., and Trejo, A. *Fluid Phase Equilib.*, **185**, 275–293(2001).
2. Arce, A., Blanco, A., Martinez-Ageitos, J. and Vidal, I. *Fluid Phase Equilib.*, **109**, 291–297 (1995).
3. Ghanadzadeh, H. and Ghanadzadeh, A. *Fluid Phase Equilib.*, **202**, 337–344 (2002).
4. Ghanadzadeh, H. and Ghanadzadeh, A. *J. Chem. Thermodynamics*, **35**, 1393–1401 (2003).
5. Briones, J. A., Mullins, J. C., and Thies, M. C. *Ind. Eng. Chem. Res.*, **33**, 151–156 (1994).
6. Dramur, U. and Tatli, B. *J. J. Chem. Eng.*, **38**, 23–25 (1993).
7. Escudero, I. and Cabezas, J. L. *J. Chem. Eng. Data*, **41**, 2–5 (1996).
8. Weast, R. C. *Handbook of Chemistry and Physics.* 17th ed., CRC Press, Boca Raton, FL, (1989–1990).
9. Escudero, I., Cabezas, J. L., and Coca, J. *J. Chem. Eng. Data*, **39**, 834–839 (1994).
10. Othmer, D. F. and Tobias, P. E. *Ind. Eng. Chem. Res.*, **34**, 690–700 (1942).

10 Some Aspects of Phase Equilibria Behavior and Verification

CONTENTS

10.1 INTRODUCTION

Liquid-liquid equilibrium data are essential for the comprehension of extraction process, and solvent capacity, solubility, selectivity, and other extraction variable can be estimated from these data. The efficient separation of organic acids and alcohols from aqueous solution is an important subject in chemical fermentation industries and many solvent have been used to improve the recovery processes [1, 2].

The phase behavior experimental data for (water + carboxylic acid + 1-octanol) and (water + carboxylic acid or alcohol +1-hexanol) systems were reported by Senol [3, 4]. Letcher and Redhi [5] studied the phase behavior of (butanenitrile + a carboxylic acid + water) systems and Bilgin et al. [6] tested the ternary system behavior of (water + propionic acid + alcohol) systems.

Certain liquid-liquid equilibrium data of the ternary systems were also investigated by non-random two liquid (NRTL) [7], universal quasi chemical theory (UNIQUAC) [8], universal functional group activity coefficient (UNIFAC) [9], Dortmund modified UNIFAC [10] models. The experimental data [3, 4] were used in order to analysis the liquid-liquid equilibrium behavior in this chapter. The experimental data [3, 4] were compared with Aspen Plus 11.1.

The main purpose of this chapter is to verify the phase equilibria behavior for ternary liquid systems, using different models. The NRTL [7] models for strongly non-ideal mixture, and especially for partially immiscible systems, the NRTL equation often provides a good representation of experimental data if the adjustable parameters obtained by enough attention using experimental data. The UNIQUAC [8] equation is applicable to a wide variety of non-electrolyte liquid mixture containing non-polar or polar fluids such as hydrocarbons, alcohols, nitriles, ketones, aldehydes, organic acid, and water including partially miscible mixture. With only two adjustable binary parameters, it cannot always represent high quality data with high accuracy, but for many typical mixtures encountered in chemical practice, UNIQUAC provides a satisfactory description.

The UNIFAC [9] activity coefficient model is an extension of the UNIQUAC model. It applies the same theory to functional groups that UNIQUAC uses for molecules. A limited number of functional groups are sufficient to form an infinite number of different molecules. Group-group interactions determined from a limited, well chosen set of experimental data are sufficient to predict activity coefficients between almost any pair of components. The UNIFAC (Fredenslund et al., 1975, 1977) can be used to predict activity coefficients for VLE. For LLE a different dataset must be used. Mixture enthalpies, derived from the activity coefficients are not accurate. This model can be applied to VLE, LLE, and enthalpies (Larsen et al., 1987). Another UNIFAC modification comes from the University of Dortmund (Germany). This modification is similar to Lyngby modified UNIFAC, but it can also predict activity coefficients at infinite dilution (Weidlich and Gmehling, 1987). The UNIFAC modification by Gmehling and coworkers (Gmehling et al., 1993, Weidlich and Gmehling, 1987), is slightly different in the combinatorial part [10].

10.2   DISCUSSION AND RESULTS

The compositions of experimental data for the ternary mixtures of (water + acetic acid +1-hexanol) and (water + acetic acid + 1-octanol) obtained at $T = 293.15K$, atmospheric pressure are shown in Tables 1 and 2 [3, 4]. The corresponding triangular diagrams for the ternary mixtures of (water + acetic acid +1-hexanol) and (water + acetic acid + 1-octanol) are presented in Figures 1, 2, 3, and 4. The NRTL, UNIQUAC, UNIFAC, and Dortmund modified UNIFAC equations of state were used in order to compare with the experimental data. The comparison between calculated and experimental compositions (water + acetic acid +1-hexanol) can be seen in Figures 1 and 2.

TABLE 1   Experimental data compositions (mass fraction) at $T = 293.15K$ [3].

Water-rich		1-hexanol-rich	
w_1	w_2	w_1	w_2
Water (1) + Acetic acid (2) + 1-hexanol (3)			
0.9922	0	0.0305	0
0.9624	0.0296	0.0383	0.0198
0.9022	0.0889	0.0514	0.0648
0.8553	0.1345	0.0705	0.1032

TABLE 1   *(Continued)*

Water-rich		1-hexanol-rich	
w_1	w_2	w_1	w_2
0.7892	0.1983	0.0919	0.1591
0.699	0.2827	0.1217	0.2345
0.648	0.3315	0.142	0.2827

The results were considered very satisfactory for Dortmund modification UNIFAC and UNIFAC models for the ternary mixtures of (water + acetic acid + 1-hexanol). However, the Dortmund modification UNIFAC equation of state is more capable to estimate the equilibrium data for the systems under study.

The Liquid-liquid equilibrium data were also estimated by UNIQUAC and NRTL methods, which have not good agreement with experimental data, however the performances of NRTL and UNIFAC models in water rich phase were satisfactory, but Dortmund modified UNIFAC and UNIQUAC models, are more efficient for predicting the equilibrium data in 1-hexanol-rich phase.

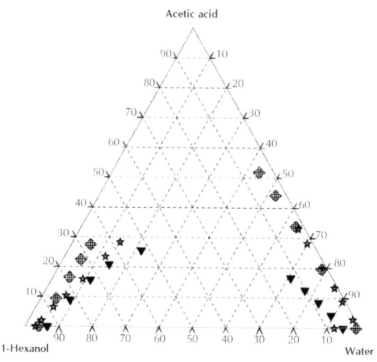

FIGURE 1   (Liquid + liquid) equilibrium data for the following systems at T = 293K: Experimental data (★), Dortmund modification UNIFAC—predicted end compositions (◈), and UNIQUAC—predicted end compositions (▼).

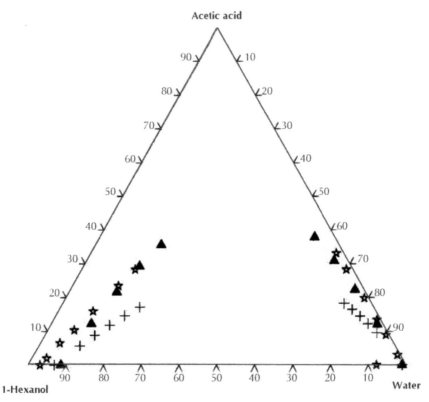

FIGURE 2 (Liquid + liquid) equilibrium data for the following systems at $T = 293K$: Experimental data (★), NRTL—predicted end compositions (+), and UNIFAC—predicted end compositions (▲).

TABLE 2 Experimental data compositions (mol fraction) at T = 293.15K [4].

Water-rich		1-octanol-rich	
X_1	X_2	X_1	X_2
Water (1) + Formic acid (2) + 1-octanol (3)			
0.9759	0.0237	0.2033	0.0982
0.9172	0.0822	0.1868	0.2814
0.848	0.1511	0.1718	0.4199
0.7783	0.2204	0.1626	0.5178
0.7143	0.2841	0.1795	0.5708

TABLE 2 *(Continued)*

Water-rich		1-octanol-rich	
X_1	X_2	X_1	X_2
0.6412	0.3566	0.2368	0.5878
0.9759	0.0237	0.2033	0.0982
0.9172	0.0822	0.1868	0.2814
0.9759	0.0237	0.2033	0.0982
0.9172	0.0822	0.1868	0.2814

The ternary mixtures (water + formic acid + 1-octanol) results have not well agree with experimental data, while UNIQUAC model shows relatively better results with respect to the other equations.

According to Figure 3, Dortmund modification UNIFAC model can predict the experimental data in 1-octanol rich phase better the UNIFAC model. However, the letter is able to follow the trend of end composition with a considerable difference.

The Dortmund modification UNIFAC, UNIFAC, and UNIQUAC equation of states, as shown in Figure 3 and Figure 4, are able to estimate the end composition with moderate accuracy, because of low solubility of 1-octanol in water rich phase.

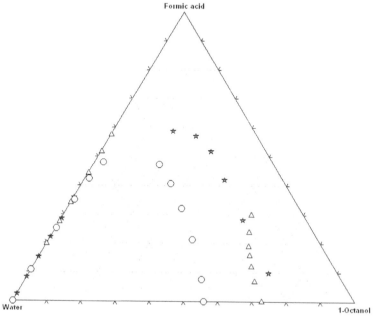

FIGURE 3 (Liquid + liquid) equilibrium data for the following systems at $T = 293K$: experimental data (★), Dortmund modification UNIFAC—predicted end compositions (◈), UNIQUAC—predicted end compositions (▼).

As seen in Figure 4, UNIFAC equation has the same behavior with Dortmund modification UNIFAC. However, there is a considerable difference in its prediction, too. The NRTL equation has also used to the ternary system under study, but its results were not acceptable, because it is not able to distinguish the water and organic phases.

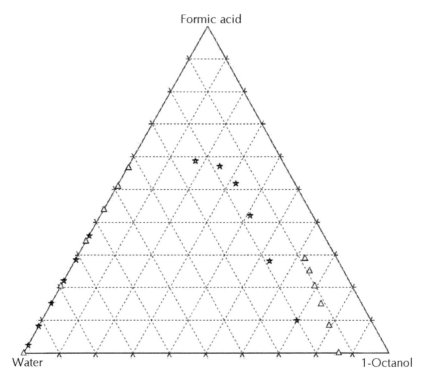

FIGURE 4 (Liquid + liquid) equilibrium data for the following systems at $T = 293$K: Experimental data (★), UNIFAC—predicted end compositions (▲).

10.4 CONCLUSION

The LLE data of the ternary mixtures of (water + acetic acid + 1-hexanol), (water + formic acid + 1-octanol) at T = 293.15K, and ambient pressure were studied with NRTL, UNIFAC, Dortmund modified UNIFAC, and UNIQUAC equations. Dortmund modified UNIFAC and UNIQUAC models showed good agreement with experimental results, in system (water + acetic acid + 1-hexanol). The UNIQUAC model presents relatively better performance in second system (water + formic acid + 1-octanol), with respect to the other equation. It must consider that, NRTL model has failed in estimating the end composition for system (water + formic acid + 1-octanol), because of simple structure. The UNIFAC and Dortmund modified UNIFAC activity coefficient models satisfactorily correlated the LLE experimental data of the studied systems are presented.

KEYWORDS

- **Non random two liquid**
- **Phase equilibria behavior**
- **Ternary mixtures**
- **Universal functional group activity coefficient**
- **Universal quasi chemical theory**

REFERENCES

1. DongChu, C., HongQi, Y., and Hao, W. Liquid–liquid equilibria of methylcyclohexane –benzene–*N*-formylmorpholine at several temperatures. *Fluid Phase Equilibria*, **255**(5), 115 (2007).
2. Letcher, T. M. and Redhi, G. G. Phase equilibria for liquid mixtures of (benzonitrile + a carboxylic acid + water) at T = 298.15K. *J. Chem. Thermodynamics.*, **33**(11), 1555 (2001).
3. Senol, A. Phase equilibria for ternary liquid systems of (water + carboxylic acid or alcohol + 1-hexanol) at T = 293.15K: modeling considerations. *J. Chem. Thermodynamics.*, **36**(7), 1007 (2004).
4. Senol, A. Liquid-liquid equilibria for the system (water + carboxylic acid + chloroform): Thermodynamic modeling. *Fluid Phase Equilibria*, **243**(5), 51 (2006).
5. Letcher, T. M. and Redhi, G. G. Phase equilibria for liquid mixtures of (butanenitrile + a carboxylic acid + water) at 298.15K. *Fluid Phase Equilibria.*, **193**(10), 123 (2002).
6. Kırbaslar, S. I., Sahin, S., and Bilgin, M. (Liquid + liquid) equilibria of (water + propionic acid + alcohol) ternary systems. *J. Chem. Thermodynamics.*, **38**(6), 1503 (2006).
7. Renon, H. and Prausnitz, J. M. Local compositions in thermodynamic excess functions for liquid mixtures. *AIChE J.*, **14**(9), 135 (1968).
8. Abrams, D. S. and Prausnitz, J. M. Local compositions in thermodynamic excess functions for liquid mixtures. *AIChE J.*, **21**, 116–128 (1975).
9. Fredenslund, A., Jones, R. L., and Prausnitz, J. M. Group-contribution estimation of activity coefficients in nonideal liquid mixtures. *AIChE J.*, **21**(12), 1086 (1975).
10. Gmehling, J., Li, J., and Schiller, M. A Modified UNIFAC Model 2. Present Parameter Matrix and Results for Different Thermodynamic Properties. *Ind. Eng. Chem. Res.*, **32**(15), 178 (1993).

11 Update on Application of Response Surface Methodology-Part I

CONTENTS

11.1 INTRODUCTION

In recent years, dendritic polymers have attracted increasing attention due to their unique chemical and physical properties. These polymers consist of three subsets namely hyperbranched polymers (HBPs), dendrigraft polymers, and dendrimers. The HBPs are highly branched, polydisperse, and 3D macromolecules synthesized from a multifunctional monomer to produce a molecule with dendritic structure [1-10].

Dendrimers are well-defined and needed a stepwise route to construct the perfectly symmetrical structure. Hence, synthesis of dendrimers is time-consuming and expensive procedures. Although HBPs are irregularly shaped and not perfectly symmetrical like dendrimers, HBPs rapidly prepared and generally synthesized by one-step process *via* polyaddition, polycondensation, radical polymerization, and so on, of AB_x type monomers and no purification steps are needed for their preparation. Therefore, HBPs are attractive materials for industrial applications due to their simple production process [1-4, 11-19]. In general, according to molecular structures and properties, HBPs represent a transition between linear polymers and perfect dendrimers. Comparison of HBPs with their linear analogues indicated that HBPs have remarkable properties, such as low melt and solution viscosity, low chain entanglement, and high solubility, as a result of the large amount of functional end groups and globular structure [10-16].

In recent years, many functional HBPs with various terminal groups such as hydroxyl, amine, carboxyl, acetoxy, and vinyl have been suggested as excellent candidate for use in drug delivery [20, 21], gene therapy vectors [21, 22], coatings [23, 24], additives [25], catalysis [26], gas separation [27], nanotechnology, supramolecular science [28], and many more.

Some examples of modification of fibers with HBPs were successful. For instance, the dyeability of modified polypropylene (PP) fibers by HBP was investigated [29]. The results showed that the incorporation of HBP prior to fiber spinning considerably improved the color strength of PP fiber with C.I. Disperse Blue 56 and has no significant effect on physical properties of the PP fibers.

The synthesized amino-terminated HBP was found to be used as salt-free dyeing auxiliary for treated cotton fabrics [30]. The washing fastness, rubbing fastness, and leveling properties of HBP treated cotton fabrics were better than untreated cotton fabrics. In the study on applying amino-terminated HBP to cotton fabric, it was demonstrated that HBP treatment on cotton fabrics has no undesirable effect on mechanical properties of fabrics. Dyeing of treated cotton fabrics with direct and reactive dyes in the absence of electrolyte showed that the color strength of treated samples were better than untreated cotton fabrics. Furthermore, application of HBP to cotton fabrics reduced UV transmission and showed good antibacterial activities [31-34].

In recent years, there has been considerable research to improve polyethylene terephthalate (PET) fabrics dyeability. It is known that PET fabrics are usually dyed using disperse dyes in the presence of a carrier or at elevated temperatures. Many attempts have been made during the last two decades to replace environmentally unfriendly carriers with non-toxic chemicals [35]. Modification of PET fabrics by attaching additives or functional groups to the polymer molecules *via* grafting and copolymerization are the common methods to enhance the dyeability of PET fabrics.

Literature review showed that there has not been a previous report regarding the treatment of amine terminated HBPs on PET fabric and study of its dyeability with acid dyes. In the most recent investigation in this field, fiber grade PET was compounded with polyesteramide HBP and dyeability of resulted samples with disperse dyes was studied [35]. The results showed that the dyeability of dyed modified samples comprised of fiber grade PET films and a HBP (Hybrane H 1500) were better than the neat PET and this was increased by increasing amount of HBP in presence or absence of a carrier. The dyeability of the samples was attributed to decrease in glass transition temperature for blended PET/HBP polymer in comparison with neat PET [29].

In the first part of our present work, novel HBP with amine terminal group was synthesized from methyl acrylate and diethylene triamine by melt polycondensation. The obtained HBP was characterized using Fourier transform infrared spectroscopy (FTIR) and nuclear magnetic resonance (NMR). In this contribution for the first time, the synthesized HBP applied to PET fabric and the dyeability of HBP treated PET fabrics and untreated PET fabric with C.I. Acid Red 114 was investigated. Dyeing of PET fabric with acid dye is very helpful and attractive object.

In the second part, a different case study was presented to show how the treatment conditions can be affected the dyeability of PET fabrics. Response surface methodology (RSM) was used to obtain a quantitative relationship between selected treatment conditions and PET fabrics dyeabiliy.

11.2 EXPERIMENTAL

11.2.1 Case 1: Synthesis, Characterization, and Application of HBP to PET Fabrics

Materials

Methyl acrylate, diethylene triamine of pure grade were obtained from Merck, Germany. All reagents used were analytically pure. The PET fabric (28×19 count/cm^2) used throughout this work and before use it was treated with a solution containing 5 g/l Na_2CO_3 and 1 g/l of a non-ionic detergent at 60°C for 30 min to remove undesired materials. Distilled water was used for the treatments and washings. Acid dye (C.I. Acid Red 114) was provided by the Ciba Ltd. (Tehran, Iran) and used to evaluate the dye absorption behavior (dyeability) of samples. The chemical structure of AR 114 is shown in Figure 1.

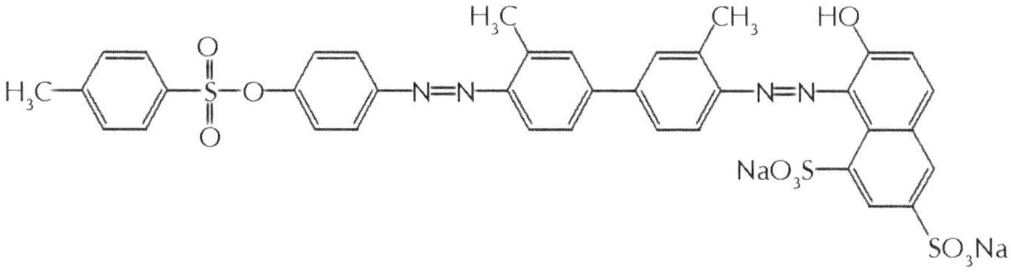

FIGURE 1   The chemical structure of C. I. Acid Red 114.

Measurements

The infrared spectra of samples were recorded by Nicolet 670 FTIR spectrophotometer in the wave number range 500–4,000 cm^{-1}. Nominal resolution for all spectra was 4cm^{-1}.

The ^{1}H and ^{13}C NMR spectra were recorded by Bruker AVANCE (500 MHz for ^{1}H and 125 MHz for ^{13}C) using DMSO-d_6 as a solvent at room temperature. All chemical shifts (δ) are expressed in ppm, the following abbreviations are used to explain the multiplicities: s = singlet, d = doublet, t = triplet, q = quartet, and m = multiplet.

X-ray diffraction (XRD) patterns of hyperbranched-treated and untreated polyester fabric were carried out with Equinox 3,000 X-ray diffractometer with Cu Kα radiation ($\lambda = 0.154$ nm) and operated at 40 kV voltage and 30 mA electric current.

The ultraviolet–visible (UV–vis) absorption spectrum of HBP was taken with a carry 100 UV–vis spectrometer.

The zeta potential of the untreated polyester fabric and the HBP treated polyester fabrics were measured by a Molvern Zetasizer 3,600 device. Fabric suspension prepared by pulling out the fibers from the fabric and cutting the fibers into lengths of approximately 0.5 mm. Then the small piece of fibers soaked in deionized water for 24 hr.

The color parameters of dyed samples were obtained under illuminant D$_{65}$ at 10°C standard observers in the visible range using Color-eye 7,000A spectrophotometer. The Kubelka-Munk single constant theory was used to calculate K/S values at the wavelength of maximum absorption (λ_{max}) for each fabric as following equation:

$$\frac{K}{S} = \frac{(1-R)^2}{2R} - \frac{(1-R_0)^2}{2R_0} \tag{1}$$

where R is the reflectance value at the wavelengths of maximum absorbance (λ_{max}), K is the absorption coefficient, and S is the scattering coefficient.

Washing fastness test was performed according to ISO/R 105/IV, Part 10. Rubbing fastness test was performed according to ISO/R 105/IV, Part 18.

11.2.2 Synthesis

Synthesis of Ab$_2$ Type Monomer

Diethylene triamine (0.5 mol, 52 ml) was added in a three-necked flask equipped with a constant voltage dropping funnel, condenser and a nitrogen inlet. The flask was placed in an ice bath and solution of methyl acrylate (0.5 mol, 43 ml) in methanol (100 ml) was added drop wise into the flask. The reaction mixture was stirred with a magnetic stirrer. Then the mixture was removed from the ice bath and left to react with a flow of nitrogen at room temperature (Scheme 1). After stirring for 4 hr, the nitrogen flow was removed and AB$_2$ type monomer was obtained.

IR v_{max} (cm^{-1}, KBr), 3,300–3,500 (N–H), 2956, 2832, 1735 (C=O).

Synthesis of Hyperbranched Polymer

The obtained light yellow and viscous mixture was transferred to an eggplant-shaped flask for an automatic rotary vacuum evaporator to remove the methanol under low pressure. Then the temperature was raised to 150°C using an oil bath and condensation reaction was carried out for 6 hr. A pale yellow viscous HBP was obtained. The ^{1}H and ^{13}C-NMR spectrum was in accordance with the proposed structure.

IR v_{max} (cm^{-1}, KBr), 3,300–3,500 (N–H), 1631 (C=O). ^{1}H NMR (ppm, DMSO-d$_6$), 4.95 (s), 3.45 (m), 2.64 (m). ^{13}C NMR (ppm, DMSO-d$_6$), 171.98, 59.86, 57.71, 40.68–39.68 (group).

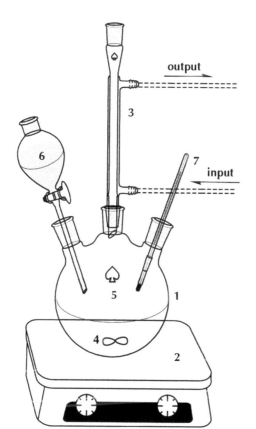

(1) three-necked flask
(2) heater stirrer
(3) condenser
(4) magnet stirrer
(5) Diethylene triamine
(6) Methyl acrylate & methanol
(7) nitrogen purge

SCHEME 1   Schematic of reactor for monomer synthesis.

PET Fabric Treatment with Hyperbranched Polymer

The PET fabrics were immersed in sodium hydroxide solution (10% w/v) at 94°C for 1 hr, using a liquor ratio of 40:1. After the alkaline treatment of fabrics, the samples were thoroughly rinsed with distilled water and neutralized with acetic acid, then rinsed and dried at room temperature. Application of HBP to original PET fabric and alkali-treated PET fabric (HPET) was carried out using exhaust method. PET samples were immersed in a HBP solution (20 g/dl) and the temperature was raised at a rate of 2.5°C/min. After 60 min, the samples were thoroughly rinsed with distilled water to remove unfixed HBP and dried at room temperature. Figure 2 shows this application profile.

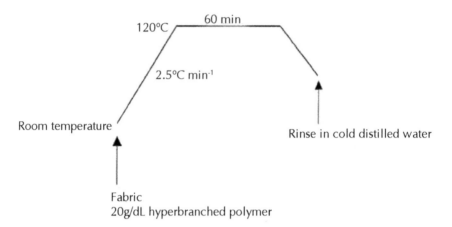

FIGURE 2 Hyperbranched polymer application profile.

The HBP treated and untreated PET fabrics were subjected to five repeated washing fastness test for 30 min using sodium carbonate (2 g/l) and detergent (5 g/l) at 60°C, as shown in Figure 3.

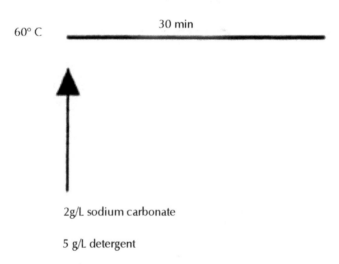

FIGURE 3 Washing fastness test profile.

Dyeing Procedure with Acid Dye

The dyeability of the hyperbranched treated PET fabric and untreated polyester fabric were examined using C.I. Acid Red 114 (AR 114). All samples were dyed competitively in the same dye bath. The treated PET fabrics were introduced into the dye bath at 40°C, then the temperature was gradually raised up to boiling point within 20 min, and then dyeing was continued for 60 min with occasional stirring. Dyeing process was carried out using a liquor ratio of 60:1. At the end of dyeing the dyed samples were rinsed with cold water, then with hot water at about 50°C and finally rinsed with tap water. Figure 4 show the dyeing profile.

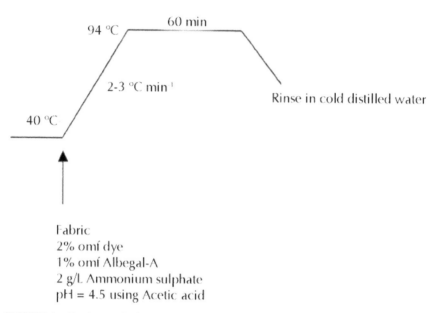

FIGURE 4   Dyeing method.

11.2.3   Case 2: Experimental Design and Optimization of Treatment Conditions

The RSM is a combination of mathematical and statistical techniques used to evaluate the relationship between a set of controllable experimental factors and observed results. This optimization process involves three major steps: (i) performing statistically designed experiments, (ii) estimating the coefficients in a mathematical model, and (iii) predicting the response and checking the adequacy of the model [36]. The RSM is used in situations where several input variables influence some output variables (responses) of the system. The main goal of RSM is to optimize the response, which is influenced by several independent variables, with minimum number of experiments. The RSM, which is a powerful and useful experimental design tool, is capable of fitting a second-order (quadratic) prediction equation for the response. Central composite design (CCD), first introduced by Box and Wilson, is the most common type of second-order designs that used in RSM and is appropriate for fitting a quadratic surface [37].

In the present study CCD was employed for the optimization of PET fabrics treated with HBP as shown in Table 1. The experiment was performed for at least three levels of each factor to fit a quadratic model. Based on preliminary experiments, HBP solution concentration (X_1), treatment temperature (X_2), and time (X_3) were determined as critical factors with significance effect on treated PET fabrics. These factors were three independent variables and chosen equally spaced, while the fabrics dyeability (K/S value) was dependent variable (response). A schematic of experimental design is shown in Figure 5. –1, 0, and 1 are coded variables corresponding to low, intermediate and high levels of each factor respectively. The actual design experiment for three independent variables listed in Table 2.

TABLE 1 Design of experiment (factors and levels).

Factor	Variable	Unit	Factor level		
			–1	**0**	**1**
X_1	Hyperbranched polymer concentration	(wt.%)	2	6	10
X_2	Temperature	(°C)	90	110	130
X_3	Time	(min)	45	60	75

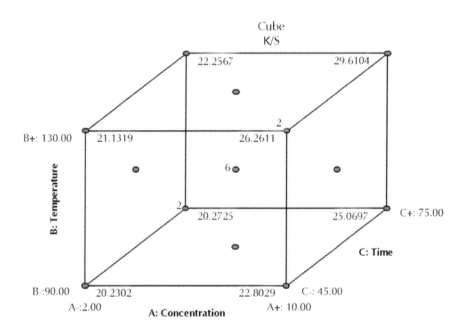

FIGURE 5 Design of experiment schematic.

The following quadratic model, which also includes the linear model, was fitted to the data presented in Table 2.

$$Y = \beta_0 + \sum_{i=1}^{k} \beta_i . x_i + \sum_{i=1}^{k} \beta_{ii} . x_i^2 + \sum \sum_{i<j=2}^{k} \beta_{ij} . x_i x_j + \varepsilon \qquad (2)$$

where, Y is the predicted response, x_i and x_j are coded variables, β_0 is constant coefficient, β_i is the linear coefficients, β_{ii} is the quadratic coefficients, β_{ij} is the second-order interaction coefficients, k is the number of factors, and ε is the approximation error.

The experimental data in Table 2 were analyzed using Design-Expert® software including analysis of variance (ANOVA). The values of coefficients for parameters $(\beta_0, \beta_i, \beta_{ii}, \beta_{ij})$ in Equation (2) and p-values are calculated by regression analysis. The coefficient of determination (R^2) has been explained by the regression model.

TABLE 2 Design of experiment (coded values).

Expt. no.	Coded value		
	X_1	X_2	X_3
1	−1	−1	1
2	−1	1	1
3	−1	0	0
4	−1	1	−1
5	−1	−1	−1
6	0	0	0
7	0	0	0
8	0	0	0
9	0	0	0
10	0	0	0
11	0	0	0
12	0	0	−1
13	0	0	1
14	0	−1	0
15	0	1	0
16	1	−1	−1
17	1	−1	1
18	1	1	1
19	1	1	−1
20	1	0	0

11.3 DISCUSSION AND RESULTS

11.3.1 Synthesis of Hyperbranched Polymer

As mentioned earlier, preparation of HBP requires two-step reaction comprise preparation of AB$_2$ type monomers (1 and 2) by Michael addition reaction of methyl acrylate and diethylene triamine and synthesis of HBP by polycondensation reaction
respectively (Scheme 2). It is well known that in HBP, the structural units include
terminal (T), dendritic (D), and linear (L) units that are shown in Scheme 2.

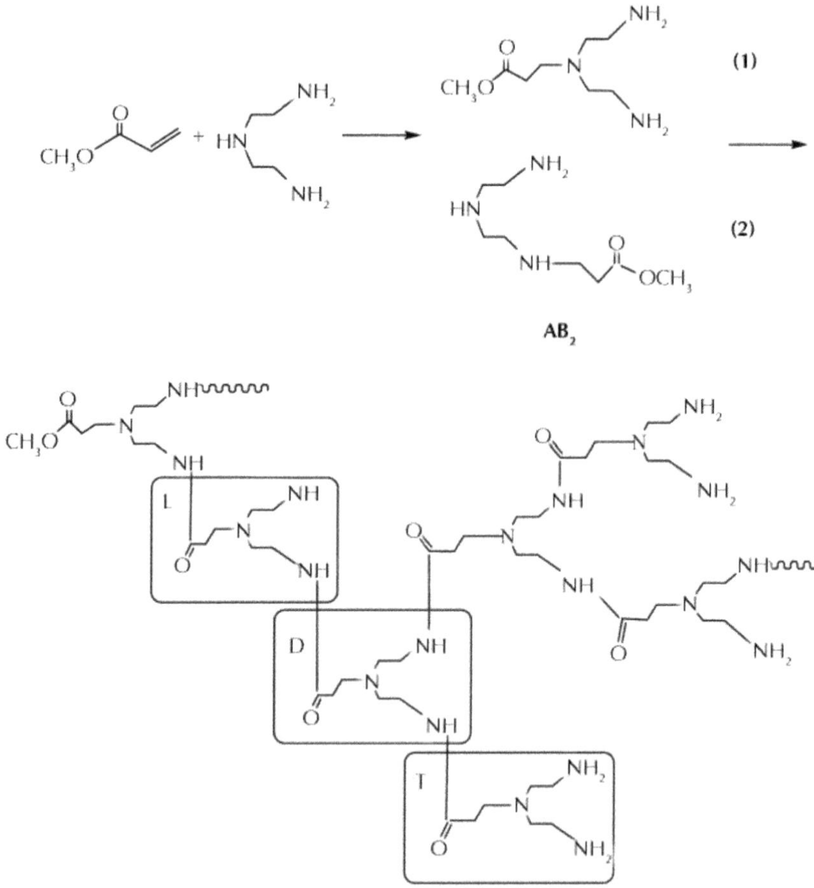

SCHEME 2   Chemical structure of amine terminated HBP.

11.3.2 FTIR Spectroscopy Characterization

HBP Analysis

To confirm the polycondensation reaction progress, AB$_2$ type monomer and HBP were
examined by FTIR spectroscopy. Spectral differences between AB$_2$ type monomer and

HBP are observed in the fingerprint region between 1,800 and 900 cm⁻¹. The absorption bands in FTIR spectra are assigned according to literature [38]. The FTIR spectra of the AB₂ type monomer and the HBP are given in Figure 6. The FTIR spectra of both monomer and polymer show the peak position at 3,300 cm⁻¹ due to N–H stretching vibration. AB₂ type monomer had absorption band at around 1,463 cm⁻¹ (bending vibration of CH₂) and bands between 1,000 and 1,400 cm⁻¹. The peaks can be observed at 1,735 cm⁻¹ corresponding to C=O stretching vibration of esters.

The FTIR spectrum of HBP is characterized by absorption at around 1,631 cm⁻¹ (C=O stretching vibration of amides) and 1,463 cm⁻¹ (bending vibration of CH₂). The absorption at 1,735 cm⁻¹ that is attributed to the C=O stretching vibration of esters is generally weak in HBP. Melt-polycondensation reaction and synthesis can be responsible for this change in the band intensity.

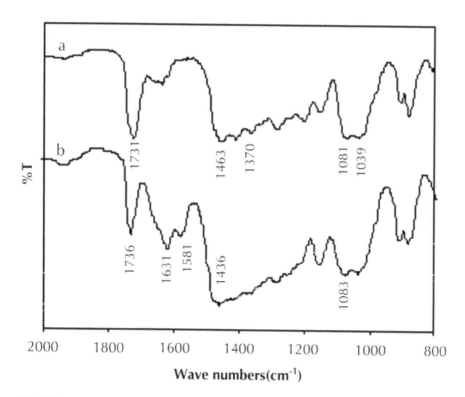

FIGURE 6   FTIR spectra of (a) AB₂ type monomer and (b) hyperbranched polymer.

Hyperbranched Treated PET Fabrics Analysis

In order to confirm the preparation process, we studied the FTIR spectroscopic data. FTIR spectra of untreated PET fabric and HBP treated PET fabric are shown in Figure 7. The FTIR of both samples shows that the peak positions are at 2,929, 2,867, 1,714,

1,238, and 1,093 cm⁻¹. The band at 1,238 and 1,093 cm⁻¹ is due to C-O stretching. While the band at 1,714 cm⁻¹ reflects the carbonyl group stretching. The characteristics absorption band of HBP at around 1,600 cm⁻¹, which reflects the N-H bending of HBP, appeared after the treatment of PET fabric.

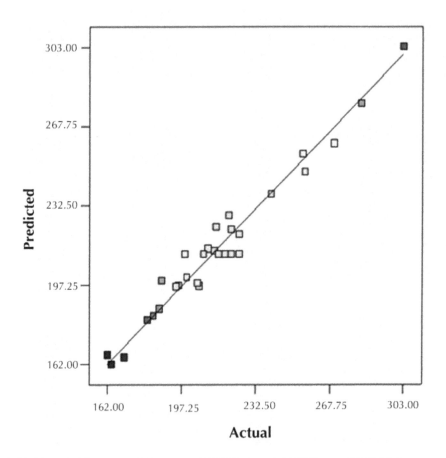

FIGURE 7   FTIR spectra of (a) untreated PET fabric and (b) HBP treated PET fabric.

11.3.3   NMR Spectroscopy Characterization
¹H NMR Analysis
The possible structure of the novel HBP was supported with NMR spectroscopy. Figure 8 shows the ¹H NMR spectrums and its assignment of amine terminated HBP. A good correlation is found between the ¹H NMR spectrum of HBP and its expected structure.

The chemical shifts ranged from δ = 2.80 – 3.70 ppm were assigned to protons of methylene. The methylene protons adjacent to the carbonyl unit give rise to a signal at δ = 2.90 ppm. The presence of the amine unit is confirmed by the resonance at δ = 2.74 ppm for the NH₂ of the amine group. The chemical shift at δ = 2.50 ppm are related to the DMSO-d₆ as a solvent [39, 40]. The proton absorption of amid group (-CONH-) was found between δ = 5.15 and 5.40 ppm with its peak at δ = 5.28 ppm [41].

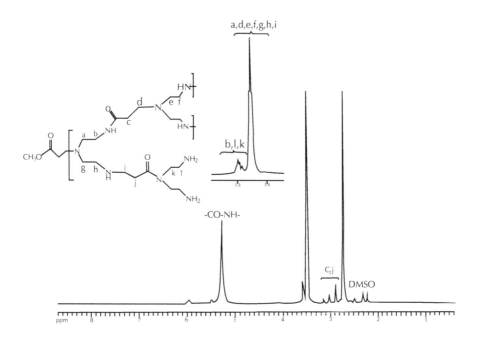

FIGURE 8   ¹H NMR spectrum of amine terminated hyperbranched polymer.

¹³C NMR Analysis

Both ¹H NMR and ¹³C NMR have been utilized to confirm the structure of HBP. The ¹³C NMR spectrum of HBP in DMSO-d₆ is shown in Figure 9. This spectrum is in conformity with the expected branched structure. The methylene carbon appears at δ = 50–60 ppm. The chemical shift at δ = 40 ppm are due to the solvent (DMSO) and the carbonyl carbon (–C=O) resonance is downfield at δ = 171.97 ppm.

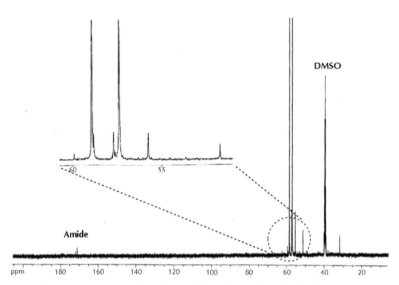

FIGURE 9 ^{13}C NMR spectrum of amine terminated hyperbranched polymer.

11.3.4   UV-vis Spectroscopy

Absorption characteristic of HBP was investigated by UV–vis spectrophotometry. The UV-vis absorption spectrum of HBP in water solution (consentration: 20 g/dl) is shown in Figure 10. It is obvious that the synthesized HBP gives strong absorption in the range of 200–350 nm. Subsequently the HBP treatment can improve the ultraviolet protection property of PET fabric.

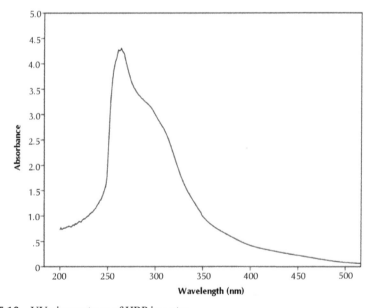

FIGURE 10   UV-vis spectrum of HBP in water.

11.3.5 Solubility Properties

Solubility properties of HBP are determined in different organic solvents included polar and nonpolar solvents. The HBP is found to be soluble in H_2O, DMSO, and DMF and insoluble in THF, xylene and toluene. Because of its highly branched structure and the nature of end functional groups, this polymer was highly soluble in polar solvents such as H_2O, DMSO, and DMF, as shown in Table 3.

TABLE 3 Solubility* properties of hyperbranched polymer.

Polymer	Solvent						
	H_2O	DMSO	DMF	THF	Toluene	Xylene	Ethanol
Hyperbranched polymer	+	+	+	−	−	−	±

The amount of polymer sample and solvents are 20 mg and 2 ml, (+)—soluble at room temperature (25°C), (±)—partially soluble, and (−)—insoluble at room temperature.

11.4 CONTACT ANGLE MEASUREMENT

11.4.1 HBP Contact Angle

Contact angle measurements were carried out by experimental equipment consisting of a camera, computer, and monitor. The contact angle of HBP aqueous solution droplet placed on a polyester plate is shown in Figure 11. Comparison of the droplet by contact angle measurements shows that the contact angle decreased dramatically with increasing concentration of HBP solution from 2.5 g/dl to 20 g/dl.

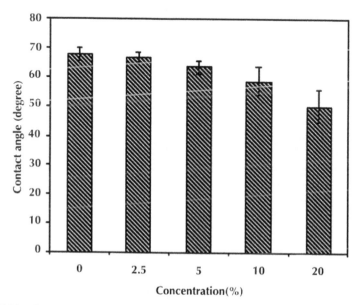

FIGURE 11   Contact angle of HBP solutions on PET plate.

11.4.2   HBP Treated PET Fabrics Contact Angle

Table 4 shows the corresponding values of the contact angle untreated and HBP treated PET fabrics. It is obvious that the alkali treatment considerably influenced the contact angle of PET fabrics and when the HPET fabric was treated with HBP, the contact angle was not change significantly. But when the PET fabric treated with HBP (PET-HBP), the contact angle was considerably decrease and HBP treated fabrics become hydrophilic. This confirms the presence of the HBP on the PET fabrics surface, which remains more hydrophilic than the untreated PET fabric.

Complete wetting of fiber surfaces is critical in all wet processes for textiles, particularly for dyeing. Uneven fiber wetting in dyeing invariably leads to uneven dye absorption. Therefore, HBP treatment cause PET fabric surface become hydrophilic and dyeing properties of PET fabrics with acid dyes become modified.

TABLE 4   Contact angle of untreated and HBP treated PET fabrics.

No.	Sample	Contact angle (°)	
		Advancing	Receding
1	PET	88 ± 6	85 ± 4
2	HPET	56 ± 6	50 ± 8
3	PET-HBP	58 ± 6	55 ± 4
4	HPET-HBP	61 ± 6	59 ± 6

11.5   X-RAY DIFFRACTION (XRD)

Figure 12 illustrate wide angle X-ray patterns of polyester samples including untreated PET, HPET, and HBP treated fabrics. All the samples have characteristic peaks around $2\theta = 17, 22$, and $25°$ corresponding to plans with miller indices (010), (110), and (100) which are in good agreement with other papers [42]. As can be seen from the Figure 12, the HBP treated PET fabrics exhibit a similar XRD pattern to untreated PET fabric, but the intensity of the $2\theta = 15–30°$ peak was decreased for PET-HBP sample, which suggested a loss of crystalline order with applying HBP.

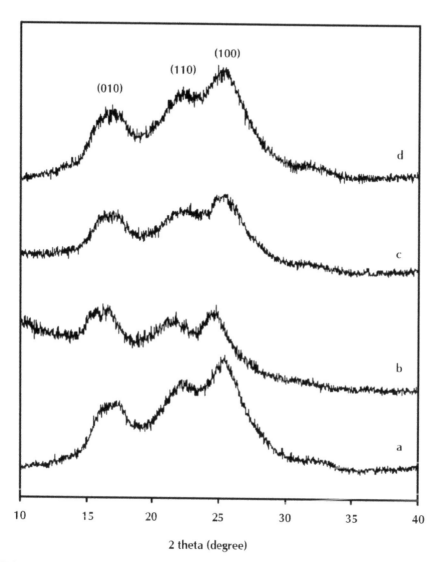

FIGURE 12   The XRD patterns of (a) PET, (b) HPET, (c) PET-HBP, and (d) HPET-HBP fabrics.

The d-spacing of crystalline part of the polymer was calculated using the Bragg Equation:

$$d = \frac{\lambda}{2\sin\theta} \tag{3}$$

where d is Bragg spacing, θ is the diffraction angle and λ is the wavelength (1.54 Å) of X-ray radiation used.

A change in the crystallite size was observed after HBP treatment. The Scherrer formula [42] was used to calculate the crystallite size as given by:

$$l_{(hkl)} = \frac{k\lambda}{\Delta_{(hkl)} \cos\theta}$$

(4)

where $l_{(hkl)}$ is the dimension of crystallite size, Δ is the full width at half maximum intensity (FWHM) of the diffraction peak and k is a constant assigned as a 1. Table 5 shows the XRD parameters for untreated and HBP treated PET fabrics.

TABLE 5   XRD parameters for untreated and HBP treated PET fabrics.

No.	Sample	Reflection index	hkl		
			010	**110**	**100**
1	PET	Crystallite size (Å)	30.9	33.6	25.8
		2θ (deg.)	17.22	22.58	25.48
		Bragg spacing (Å)	5.14	3.93	3.49
2	HPET	Crystallite size (Å)	25.6	33.6	30.5
		2θ (deg.)	16.25	21.57	24.81
		Bragg spacing (Å)	5.45	4.11	3.59
3	PET-HBP	Crystallite size (Å)	25.3	29.8	25.3
		2θ (deg.)	16.83	22.41	25.48
		Bragg spacing (Å)	5.26	3.96	3.49
4	HPET-HBP	Crystallite size (Å)	26.0	33.9	25.4
		2θ (deg.)	16.86	22.28	25.39
		Bragg spacing (Å)	5.25	3.99	3.50

Table 5 shows that the d-spacing of crystalline part of HBP treated samples are higher than untreated PET fabrics.

The broadening in the peak and change in the crystallite size due to treatment HBP–PET fabrics may be due to formation of disordered system in the HBP treated fabric.

11.6   ZETA POTENTIAL

The surface charge of samples has been investigated by zeta potential measurement to understand how HBP act on the PET fabrics. The zeta potential values of the untreated

(HPET) and the HBP treated PET fabrics (HPET-HBP) as a function of the pH of the liquid phase are shown in Figure 13. From the results, it can be seen that the zeta potentials for the untreated PET fabrics over the entire pH range are negative. While the HBP treated PET fabrics exhibited a positive charge on the surface at a low pH. This is probably attributed to the presence of numerous terminal primary amino groups on HBP treated PET fabrics, that will protonate in the liquid phase and give rise to positive charge at lower pH values. It implies that HBP enhances the adsorption of acid dye on PET fabrics.

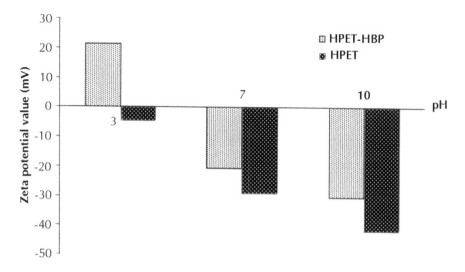

FIGURE 13   Zeta potential values of the untreated (HPET) and HBP treated PET fabrics (HPET-HBP).

11.7   DYEING PROPERTIES OF HBP TREATED POLYESTER FABRIC

The effect of HBP treatment on dye absorbance behavior (dyeability) of PET fabrics is evaluated by measuring its optical properties. The reflectance spectra of HBP treated and untreated PET fabrics are shown in Figure 14 for dyed samples. As shown in Figure 14 the reflectance spectrum of HBP treated PET fabric is less than other sample. Subsequently, the treatment of polyester fabrics by HBP significantly changes the reflectance of samples.

The CIELAB color parameter (L*, a*, b*, C*, and h°) of HBP treated and untreated PET fabrics are shown in Table 6. As shown in this table, the lightness (L*) of untreated polyester fabric is more than other sample. Also the hue (h°) of HBP treated PET fabric is more than untreated sample. Subsequently, the HBP treatment has significant change on color parameters of PET fabrics.

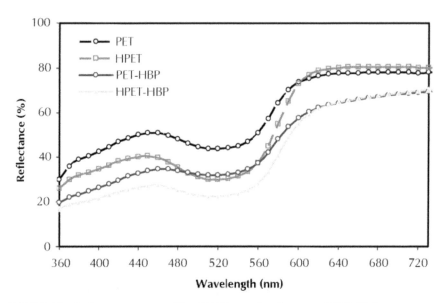

FIGURE 14   Reflectance spectra of dyed HBP treated and untreated PET fabrics.

TABLE 6   Color parameters of dyed samples at 2% omf acid dye.

No.	Samples	L*	a*	b*	C*	h°
1	PET	78.951	18.079	5.954	19.034	0.318
2	HPET	72.780	30.075	7.577	31.015	0.247
3	PET-HBP	70.291	18.432	10.525	21.225	0.519
4	HPET-HBP	64.861	27.547	11.558	29.873	0.397

The color strength (K/S) results of the HBP treated PET fabric and untreated PET fabric dyed with 2% omf Acid Red 114 using a competitive dyeing method, are shown in Figure 15. Higher K/S values indicate greater dye uptake and higher color strengths. It is clear that the color strength of the HBP treated PET fabric was very higher than that of the corresponding untreated PET fabric. The increase in the color strength may be because of the positively charged amino groups in the HBP treated PET fabric.

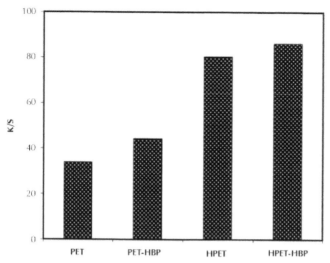

FIGURE 15   Effect of fabric treatment on color strength achieved using 2% omf C.I. Acid Red 114.

11.8   FASTNESS PROPERTIES

11.8.1   HBP Treatment Fastness

The treated samples were subjected to five repeated washing fastness test at 60°C. Figure 16 shows the washing fastness result of the undyed HBP treated and untreated PET fabric and its effect on dyeability of PET fabrics. It is obvious that the extent of color strength change was relatively small for treated and untreated PET fabrics.

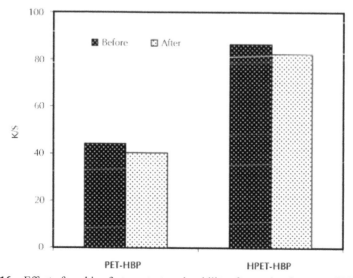

FIGURE 16   Effect of washing fastness test on dyeability of treated and untreated PET fabrics.

11.8.2 Dyeing Fastness

Table 7 shows the washing fastness and rubbing fastness of untreated PET fabrics (PET), hydrolyzed PET fabrics (HPET) and HBP treated PET fabrics (PET-HBP and HPET-HBP) dyed with acid dye. Fastness results show that the rubbing fastness of treated samples were good: the washing fastness of all sample were, however, poor, and unaffected by treatment with HBP.

TABLE 7   Fastness properties of HBP treated and untreated PET fabrics.

No.	Samples	Washing fastness*	Rubbing fastness**	
			Dry	**Wet**
1	PET	1	4	3–4
2	HPET	1	5	4–5
3	PET-HBP	1	4–5	4
4	HPET-HBP	1	4–5	4

* With gray scale
** With stain scale

11.9   THE ANALYSIS OF VARIANCE (ANOVA) AND OPTIMIZATION

All twenty experimental runs of CCD were performed in accordance with Table 8.

TABLE 8   The actual design of experiments and response.

Un-coded value			Response
HBP concentration, (wt.%)	Temperature, (°C)	Time, (min)	K/S value
2	90	75	20.38
2	130	75	21.57
2	110	60	19.86
2	130	45	21.09
2	90	45	19.78
6	110	60	22.68
6	110	60	22.18
6	110	60	22.67
6	110	60	22.97
6	110	60	22.64
6	110	60	22.26

TABLE 8 *(Continued)*

Un-coded value			Response
HBP concentration, (wt.%)	Temperature, (°C)	Time, (min)	K/S value
6	110	45	22.59
6	110	75	24.47
6	90	60	22.58
6	130	60	26.07
10	90	45	23.35
10	90	75	24.97
10	130	75	29.92
10	130	45	26.01
10	110	60	23.23

A significance level of 5% was selected: that is, statistical conclusions may be assessed with 95% confidence. In this significance level, the factor has significant impact on fabrics dyeability if the P-value is less than 0.05. And when P-value is greater than 0.05, it is concluded the factor has no significant impact on fabrics dyeability. The results of ANOVA are shown in Table 9. The regression equation was obtained from the ANOVA.

TABLE 9 Analysis of variance for response surface.

Source	Sum of squares	F-value	P-value
Model	104.71	25.98	< 0.0001
X_1-concentration	61.58	137.49	< 0.0001
X_2- temperature	18.51	41.33	< 0.0001
X_3-time	7.19	16.05	0.0025
$X_1 X_2$	3.27	7.30	0.0223
$X_1 X_3$	2.48	5.52	0.0406
$X_2 X_3$	0.59	1.31	0.2794
X_1^2	4.59	10.24	0.0095

TABLE 9   *(Continued)*

Source	Sum of squares	F-value	P-value
X_2^2	6.10	13.62	0.0042
X_3^2	1.31	2.93	0.1177
Lack of Fit	3.41	31.17	Not significant
Std.dev : 0.67	R^2 : 0.959	Adj- R^2 : 0.922	

$$K/S = +22.57 + 2.48X_1 + 1.36X_2 + 0.85X_3 + 0.64X_1X_2 + 0.56X_1X_3$$
$$+ 0.27X_2X_3 - 1.29X_1^2 + 1.49X_2^2 + 0.69X_3^2 \tag{5}$$

From the p-values presented in Table 3, it can be concluded that p-value of term X_3 is greater than the p-values for terms X_1 and X_2. Also p-value of term X_3^2 and X_2X_3 is much greater than the significance level of 0.05. And the interaction between temperature and time is not significant. But p-values for terms related to X_1 and X_2 are less than 0.05. Therefore, HBP concentration and treatment temperature have a significant impact on the HBP treated fabrics dyeability.

Because terms related to treatment time have no significant impact on samples K/S values, we removed term X_3^2 and X_2X_3 and fitted the equation by regression analysis again. The fitted equation in coded unit is given by:

$$K/S = +22.65 + 2.48X_1 + 1.36X_2 + 0.85X_3$$
$$+ 0.64X_1X_2 + 0.56X_1X_3 - 1.03X_1^2 + 1.75X_2^2 \tag{6}$$

Now, All the p-values are less than the significance level of 0.05.

The effects of three variables on K/S value of dyed samples are shown in Figure 17. Figure 17(a) shows K/S value of dyed samples at different HBP concentration and temperature for constant treatment time. The surface plot shows that at any given temperature the K/S value of samples increase with increasing the HBP concentration. Maximum K/S value was observed at high HBP concentration and high temperature. Figure 17(b) shows the response surface plot of interactions between HBP concentration and treatment time. It can be seen the increase in K/S value with increase in concentration at any given time. Moreover, it should be noted that at higher temperature, the increase in concentration and time result the higher K/S value.

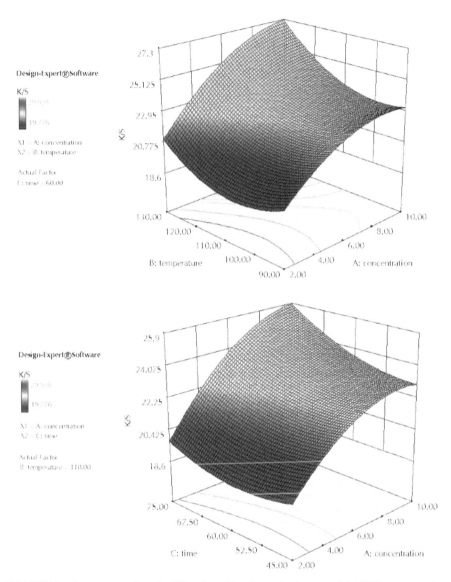

FIGURE 17   Response surface for K/S value of dyed samples in term of: (a) HBP concentration and treatment temperature and (b) HBP concentration and treatment time.

The actual and the predicted K/S value of samples plot with correlation coefficient of 0.91, is shown in Figure 18. Actual values are the measured response data for a particular run, and the predicted values evaluated from the model. It can be observed that experimental values are in good agreement with the predicted values. The optimal conditions to obtain the maximum dyeability (K/S value) efficiency are shown in Table 10.

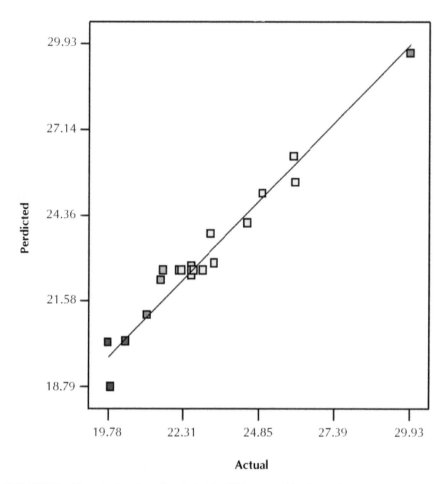

FIGURE 18   The actual and predicted plot for K/S value of dyed samples.

TABLE 10   Optimum values of the treatment conditions for maximum dyeability efficiency.

HBP concentration (wt.%)	Temperature (°C)	Time (min)	Predicted K/S	Experimental K/S	Desirability
10	130	75	29.25	29.93	0.93

11.10   CONCLUSION

In this study, a hyperbranched, functional and water-soluble polymer was successfully synthesized by melt polycondensation reaction of methyl acrylate and diethylene
triamine. The FTIR, ^{1}H and ^{13}C NMR spectroscopy measurements of HBP indicated
that this polymer comprised terminal amine group and branching was occurred. The
solubility properties of resulting HBP indicated that this polymer is readily soluble in

polar solvent such as water, DMSO, and DMF. The HBP was applied to alkali-treated PET fabric using exhaust method. Using Bragg's equation it was found that d-spacing increase after applying HBP to PET fabrics. The study of dyeability of treated samples with C.I. Acid Red 114 indicated that the color strength of HBP treated PET fabrics is more than untreated PET fabric due to the presence of terminal primary amino groups in the molecular structure of the HBP, that will protonate in the liquid phase and give rise to positive charge at lower pH values, as shown by zeta potential measurement. Samples contact angles tends to decrease with the HBP treatment, and the hydrophobic untreated PET fabrics surface becomes hydrophilic. This confirms the presence of the HBP on the PET fabrics surface and its effect on hydrophilic behavior of PET fabrics. It is noteworthy that in our attempt PET fabric dyed with acid dye effectively using HBP. Compared with previously reported article, our attempt has an innovation in PET dyeing with acid dyes.

The concentration, temperature, and time are the most important factors for HBP treatment to PET fabrics and dyeability efficiency. The RSM was successfully applied to find out the optimum level of the factors using CCD. The optimum concentration of HBP, temperature, and time were found to be 10 wt.%, 130°C and 75 min, respectively, for maximum K/S value of HBP treated samples.

KEYWORDS

- **Central composite design**
- **Fourier transform infrared spectroscopy**
- **Hyperbranched polymers**
- **Nuclear magnetic resonance**
- **Polypropylene**
- **Response surface methodology**

REFERENCES

1. Jikei, M. and Kakimoto, M. Hyperbranched polymers: A promising new class of materials. *Progress in Polymer Science*, **26**, 1233–1285 (2001).
2. Gao, C. and Yan, D. Hyperbranched polymers: From synthesis to applications. *Progress in Polymer Science*, **29**, 183–275 (2004).
3. Voit, B. I. and Lederer, A. Hyperbranched and highly branched polymer architectures synthetic strategies and major characterization aspects. *Chem. Rev.*, **109**, 5924–5973 (2009).
4. Kumar, A. and Meijer, E. W. Novel hyperbranched polymer based on urea linkages. *Chem. Commun.*, 1629–1630 (1998).
5. Grabchev, I., Petkov, C., and Bojinov, V. Infrared spectral characterization of poly(amidoamine) dendrimers peripherally modified with 1,8-naphthalimides. *Dyes and Pigments*, **62**, 229–234 (2004).
6. Qing-Hua, C., Rong-Guo, C., Li-Ren, X., Qing-Rong, Q., and Wen-Gong, Z. Hyperbranched poly (amide-ester) mildly synthesized and its characterization. Chinese *J. Struct. Chem.*, **27**, 877–883 (2008).
7. Kou, Y., Wan, A., Tong, S., Wang, L., and Tang, J. Preparation, characterization, and modification of hyperbranched polyester-amide with core molecules. *Reactive and Functional Polymers*, **67**, 955–965 (2007).

8. Schmaljohann, D. P., Pötschke, P., Hässler, R., Voit, B. I., Froehling, P. E., Mostert, B., and Loontjens, J. A. Blends of amphiphilic, hyperbranched polyesters and different polyolefins. *Macromolecules*, **32**, 6333–6339 (1999).

9. Kim, Y. H. Hyperbranched polymers 10 years after. *Journal of Polymer Science: Part A: Polymer Chemistry*, **36**, 1685–1698 (1998).

10. Liu, G. and Zhao, M. Non-isothermal crystallization kinetics of AB_3 hyper-branched polymer/polypropylene blends. *Iranian Polymer Journal*, **18**, 329–338 (2009).

11. Seiler, M. Hyperbranched polymers: Phase behavior and new applications in the field of chemical engineering. *Fluid Phase Equilibria*, **241**, 155–174 (2006).

12. Yates, C. R. and Hayes, W. Synthesis and applications of hyperbranched polymers. *European Polymer Journal*, **40**, 1257–1281 (2004).

13. Voit, B. New developments in hyperbranched polymers. *Journal of Polymer Science: Part A: Polymer Chemistry*, **38**, 2505–2525 (2000).

14. Nasar, A. S., Jikei, M., and Kakimoto, M. Synthesis and properties of polyurethane elastomers crosslinked with amine-terminated AB2-type hyperbranched polyamides. *European Polymer Journal*, **39**, 1201–1208 (2003).

15. Froehling, P. E. Dendrimers and dyes-a review. *Dyes and pigments*, **48**, 187–195 (2001).

16. Jikei, M., Fujii, K., and Kakimoto, M. Synthesis and characterization of hyperbranched aromatic polyamide copolymers prepared from AB_2 and AB monomers. *Macromol. Symp.*, **199**, 223–232 (2003).

17. Radke, W., Litvinenko, G., and Müller, A. H. E. Effect of core-forming molecules on molecular weight distribution and degree of branching in the synthesis of hyperbranched polymers. *Macromolecules*, **31**, 239–248 (1998).

18. Maier, G., Zech, C., Voit, B., and Komber, H. An approach to hyperbranched polymers with a degree of branching of 100%. *Macromol. Chem. Phys.*, **199**, 2655–2664 (1998).

19. Voit, B., Beyerlein, D., Eichhorn, K., Grundke, K., Schmaljohann, D., and Loontjens, T. Functional hyper-branched polyesters for application in blends, coations, and thin films. *Chem. Eng. Technol.*, **25**, 704–707 (2002).

20. Gao, C., Xu, Y., Yan, D., and Chen W. Water-soluble degradable hyperbranched polyesters: novel candidates for drug delivery. *Biomacromolecules*, **4**, 704–712 (2003).

21. Paleos, C. M., Tsiourvas, D., and Sideratou, Z. Molecular engineering of dendritic polymers and their application as drug and gene delivery systems. *Molecular pharmaceutics*, **4**, 169–188 (2007).

22. Wu, D., Liu, Y., Jiang, X., He, C., Goh, S. H., and Leong, K. W. Hyperbranched poly (amino ester)s with different terminal amine groups for DNA delivery. *Biomacromolecules*, **7**, 1879–1883 (2006).

23. Rolf, A. T. M. van Benthem. Novel hyperbranched resins for coating applications. *Progress in Organic Coatings*, **40**, 203–214 (2000).

24. Tang, W., Huang, Y., Meng, W., and Qing, F. Synthesis of fluorinated hyperbranched polymers capable as highly hydrophobic and oleophobic coating materials. *European Polymer Journal*, **46**, 506–518 (2010).

25. Zhang, W., Zhang, Y., and Chen, Y. Modified brittle poly (lactic acid) by biodegradable hyperbranched poly (ester amide). *Iranian Polymer Journal*, **17**, 891–898 (2008).

26. Astruc, D. and Chardac, F. Dendritic catalysts and dendrimers in catalysis. *Chem. Rev.*, **101**, 2991–3023 (2001).

27. Fang, J., Kita, H., and Okamoto, K. Hyperbranched polyimides for gas separation applications. Synthesis and Characterization. *Macromolecules*, **33**, 4639–4646 (2000).

28. Liu, C., Gao, C., and Yan, D. Synergistic supramolecular encapsulation of amphiphilic hyperbranched polymer to dyes. *Macromolecules*, **39**, 8102–8111 (2006).

29. Burkinshaw, S. M., Froehling, P. E., and Mignanelli M. The effect of hyperbranched polymers on the dyeing of polypropylene fibers. *Dyes and Pigments*, **53**, 229–235 (2002).

30. Zhang, F., Chen, Y., Lin, H., and Lu, Y. Synthesis of an amino-terminated hyperbranched polymer and its application in reactive dyeing on cotton as a salt-free dyeing auxiliary. *Coloration Technology*, **123**, 351–357 (2007).

31. Zhang, F., Chen, Y., Lin, H., Wang, H., and Zhao, B. HBP-NH$_2$ grafted cotton fiber: Preparation and salt-free dyeing properties. *Carbohydrate Polymers*, **74**, 250–256 (2008).

32. Zhang, F., Chen, Y. Y., Lin, H., and Zhang, D. S. Performance of cotton fabric treated with an amino-terminated hyperbranched polymer. *Fibers and Polymers*, **9**, 515–520 (2008).

33. Zhang, F., Chen, Y., Ling, H., and Zhang, D. Synthesis of HBP-HTC and its application to cotton fabric as an antimicrobial auxiliary. *Fibers and Polymers*, **10**, 141–147 (2009).

34. Zhang, F., Zhang, D., Chen, Y., and Lin, H. The antimicrobial activity of the cotton fabric grafted with an amino-terminated hyperbranched polymer. *Cellulose*, **16**, 281–288 (2009).

35. Khatibzadeh, M., Mohseni, M., and Moradian, S. Compounding fiber grade polyethylene terephthalate with a hyperbranched additive and studying its dyeability with a disperse dye. *Coloration Technology*, **126**, 269–274 (2010).

36. Murugesan, K., Dhamija, A., Nam, I., Kim, Y., and Chang, Y. Decolourization of reactive black five by laccase: Optimization by response surface methodology. *Dyes and Pigments*, **75**, 176–184 (2007).

37. Myers, R. H., Montgomery, D. C., and Anderson-cook, C. M. *Response surface methodology: Process and product optimization using designed experiments 3rd ed.* John Wiley and Sons, USA (2009).

38. Pavia, D. L., Lampman, G. M., Kriz, G. S., and Vyvyan, J. R. *Introduction to spectroscopy 4th ed.* Brooks/Cole, US (2001).

39. Tang, W., Huang, Y., Meng, W., and Qing, F. Synthesis of fluorinated hyperbranched polymers capable as highly hydrophobic and oleophobic coating materials. *European Polymer Journal*, **46**, 506–518 (2010).

40. Reul, R., Nguyen, J., and Kissel, T. Amine-modified hyperbranched polyesters as non-toxic, biodegradable gene delivery systems. *Biomaterials*, **30**, 5815–5824 (2009).

41. Pan, B., Cui, D., Gao, F., and He, R. Growth of multi-amine terminated poly (amidoamine) dendrimers on the surface of carbon nanotubes. *Nanotechnology*, **17**, 2483–2489 (2006).

42. Youssefi, M., Morshed, M., and Kish, M. H. Crystalline structure of poly (ethylene terephthalate) filaments. *Journal of applied polymer science*, **106**, 2703–2709 (2007).

12 Update on Application of Response Surface Methodology–Part II

CONTENTS

12.1 INTRODUCTION

In recent years, electrospinning as a simple and effective method for preparation of nanofibers materials have attracted increasing attention [1]. Electrospinning process, unlike the conventional fiber spinning systems (melt spinning, wet spinning, etc.), uses electric field force instead of mechanical force to draw and stretch a polymer jet [2]. This method provided nonwoven mat with individual fiber diameter ranged from micrometer to nanometer. Due to their high specific surface area, high porosity, and small pore size, the unique nanofibers have been suggested as excellent candidate for many applications including filtration [3], multifunctional membranes [4], biomedical agents [5], tissue engineering scaffolds [6, 7], wound dressings [8], full cell [9], and protective clothing [10].

The typical electrospinning apparatus consists of three components to fulfill the process including syringe filled with a polymer solution, a high voltage supplier to provide the required electric force for stretching the liquid jet and a grounded collection plate to hold the nanofibers mat. In electrospinning, the electrical forces applied between the nozzle and collector draws the polymer solution toward the collection plate as a jet. Usually, voltages range from 5 to 30 kV, sufficient to overcome the surface tension forces of the polymer solution. During the jet movement to the collector, the solvent evaporates and dry fibers randomly deposited on the surface of a collector [11-16]. A schematic representation of electrospinning is shown in Figure 1.

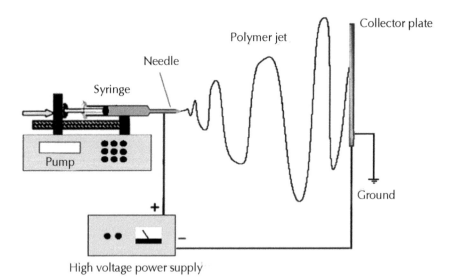

FIGURE 1   Schematic of electrospinning set up.

The morphology and the diameter of the electrospun nanofibers can be affected by many electrospinning parameters including solution properties (the concentration, liquid viscosity, surface tension, and dielectric properties of the polymer solution), processing conditions (applied voltage, volume flow rate, tip to collector distance, and the strength of the applied electric field), and ambient conditions (temperature, atmospheric pressure, and humidity) [17-20].

The wettability of solid surfaces is a very important property of surface chemistry, which is controlled by both the chemical composition and the geometrical microstructure of surface [21-23]. When a liquid droplet contacts a solid surface, it will spread or remain as droplet with the formation of angle between the liquid and solid phases. Contact angle (CA) measurements are widely used to characterize the wettability of solid surface. Surface with a water CA greater than 150° is usually called superhydrophobic surface. On the other hand, when the CA is lower than 5°, it is called superhydrophilic surface. Fabrication of these surfaces has attracted considerable interest for both fundamental research and practical studies [23-25].

In this work, we investigate the effect of four electrospinning parameters (solution concentration, applied voltage, tip to collector distance, and volume flow rate) on the average fiber diameter (AFD) and CA of electrospun PAN nanofibers mat. The aim of the present study is to establish quantitative relationship between electrospinning parameters and AFD and CA of electrospun fiber mat by response surface methodology (RSM).

12.2 EXPERIMENTAL

12.2.1 Materials

Polyacrylonitrile (PAN) powder was purchased from Polyacryle Co. (Iran). The weight average molecular weight (M_w) of PAN was approximately 100,000 g/mol. The solvent N-N, dimethylformamide (DMF) was obtained from Merck Co. (Germany). These chemicals were used as received.

12.2.2 Electrospinning

In our experiment, the PAN powder was dissolved in DMF and gently stirred for 24 hr at 50°C. Therefore, homogenous PAN/DMF solution was prepared in different concentration ranged from 10 to 14 wt.%.

Electrospinning was set up in a horizontal configuration. The electrospinning apparatus consisted of 5 ml plastic syringe connected to a syringe pump and a rectangular grounded collector (aluminum sheet). A high voltage power supply (capable to produce 0–40 kV) was used to apply a proper potential to the metal needle. It should be noted that all electrospinning were carried out at room temperature.

12.2.3 Characterization

The morphology of the gold sputtered electrospun fibers were observed by scanning electron microscope (SEM, Philips XL-30). The AFD and distribution was determined from selected SEM image by measuring at least fifty random fibers. The wettability of electrospun fiber mat was determined by CA angle measurement. The CA measurements were carried out using specially arranged microscope equipped with camera and PCTV vision software as shown in Figure 2. The droplet used was distilled water and was 1 μl in volume. The CA experiments were carried out at room temperature and were repeated five times. All CAs measured within 20 s of placement of the water droplet on the electrospun fiber mat.

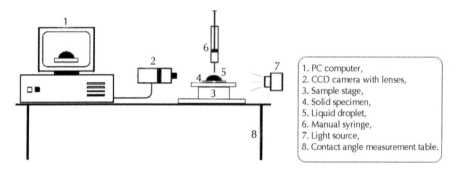

1. PC computer,
2. CCD camera with lenses,
3. Sample stage,
4. Solid specimen,
5. Liquid droplet,
6. Manual syringe,
7. Light source,
8. Contact angle measurement table.

FIGURE 2   Schematic of contact angle measurement set up.

12.2.4   Experimental Design

The RSM is a combination of mathematical and statistical techniques used to evaluate the relationship between a set of controllable experimental factors and observed results. This optimization process is used in situations where several input variables influence some output variables (responses) of the system. The main goal of RSM is to optimize the response, which is influenced by several independent variables, with minimum number of experiments. Central composite design (CCD) is the most common type of second-order designs that used in RSM and is appropriate for fitting a quadratic surface [26, 27].

The CCD was employed to establish relationships between four electrospinning parameters and two responses including the AFD and the CA of electrospun fiber mat. The experiment was performed for at least three levels of each factor to fit a quadratic model. Based on preliminary experiments, polymer solution concentration (X_1), applied voltage (X_2), tip to collector distance (X_3), and volume flow rate (X_4) were determined as critical factors with significance effect on AFD and CA of electrospun fiber mat. These factors were four independent variables and chosen equally spaced, while the AFD and the CA of electrospun fiber mat were dependent variables (responses). The values of -1, 0, and 1 are coded variables corresponding to low, intermediate, and high levels of each factor respectively. The experimental parameters and their levels for four independent variables are shown in Table 1.

TABLE 1   Design of experiment (factors and levels).

Factor	Variable	Unit	Factor level		
			−1	0	1
X_1	Solution concentration	(wt.%)	10	12	14
X_2	Applied voltage	(kV)	14	18	22
X_3	Tip to collector distance	(cm)	10	15	20
X_4	Volume flow rate	(ml/hr)	2	2.5	3

The following quadratic model, which also includes the linear model, was fitted to the data.

$$Y = \beta_0 + \sum_{i=1}^{k} \beta_i . x_i + \sum_{i=1}^{k} \beta_{ii} . x_i^2 + \sum \sum_{i<j=2}^{k} \beta_{ij} . x_i x_j + \varepsilon \tag{1}$$

where, Y is the predicted response, x_i and x_j are coded variables, β_0 is constant coefficient, β_i is the linear coefficients, β_{ii} is the quadratic coefficients, β_{ij} is the second-order interaction coefficients, k is the number of factors, and ε is the approximation error [26, 27].

The experimental data were analyzed using Design-Expert® software including analysis of variance (ANOVA). The values of coefficients for parameters ($\beta_0, \beta_i, \beta_{ii}, \beta_{ij}$) in Equation (1), p-values, the determination coefficient (R^2) and adjusted determination coefficient (R_{adj}^2) were calculated by regression analysis.

12.3 DISCUSSION AND RESULTS

12.3.1 Morphological Analysis of Nanofibers

The PAN solution in DMF were electrospun under different conditions, including various PAN solution concentrations, applied voltages, volume flow rates, and tip to collector distances, to study the effect of electrospinning parameters on the morphology and diameter of electrospun nanofibers.

Figure 3 shows the SEM images and fiber diameter distributions of electrospun fibers in different solution concentration as one of the most effective parameters to control the fiber morphology. As observed in Figure 3, the AFD increased with increasing concentration. It was suggested that the higher solution concentration would have more polymer chain entanglements and less chain mobility. This causes the hard jet extension and disruption during electrospinning process and producing thicker fibers.

The SEM image and corresponding fiber diameter distribution of electrospun nanofibers in different applied voltage are shown in Figure 4. It is obvious that increasing the applied voltage cause an increase followed by a decrease in electrospun fiber diameter. As demonstrated [17, 30], increasing the applied voltage may decrease, increase, or may not change the fiber diameter. In one hand, increasing the applied voltage will increase the electric field strength and higher electrostatic repulsive force on the jet, favoring the thinner fiber formation. On the other hand, more surface charge will introduce on the jet and the solution will be removed more quickly from the tip of needle. As a result, the AFD will be increased [29, 30].

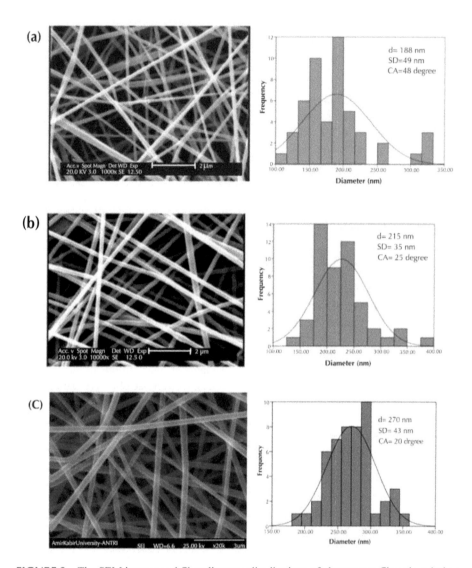

FIGURE 3   The SEM images and fiber diameter distributions of electrospun fibers in solution concentration of (a) 10 wt.%, (b) 12 wt.%, and (c) 14 wt.%.

Figure 5 represents the SEM image and fiber diameter distribution of electrospun nanofibers in different spinning distance. It can be seen that the AFD decreased with increasing tip to collector distance. Because of the longer spinning distance could give more time for the solvent to evaporate, increasing the spinning distance will decrease fiber diameter [30, 31].

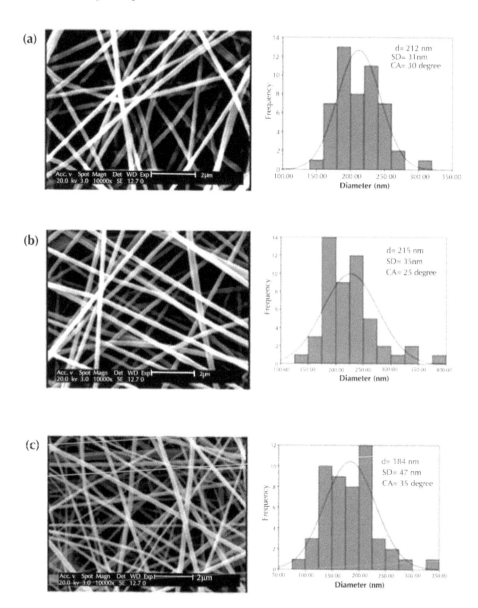

FIGURE 4   The SEM images and fiber diameter distributions of electrospun fibers in applied voltage of (a) 14 kV, (b) 18 kV, and (c) 22 kV.

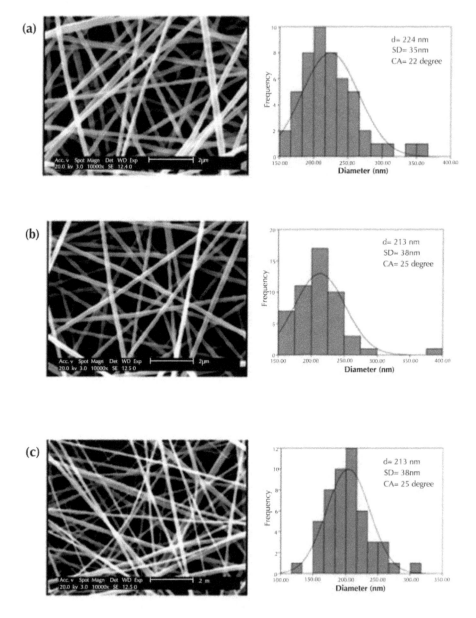

FIGURE 5 The SEM images and fiber diameter distributions of electrospun fibers in tip to collector distance of (a) 10 cm, (b) 15 cm, and (c) 20 cm.

The SEM image and fiber diameter distribution of electrospun nanofibers in different volume flow rate are illustrated in Figure 6. It is clear that increasing the volume flow rate cause an increase in AFD. Ideally, the volume flow rate must be compatible with the amount of solution removed from the tip of the needle. At low volume flow rates, solvent would have sufficient time to evaporate and thinner fibers were produced, but at high volume flow rate, excess amount of solution fed to the tip of needle and thicker fibers result [28-31].

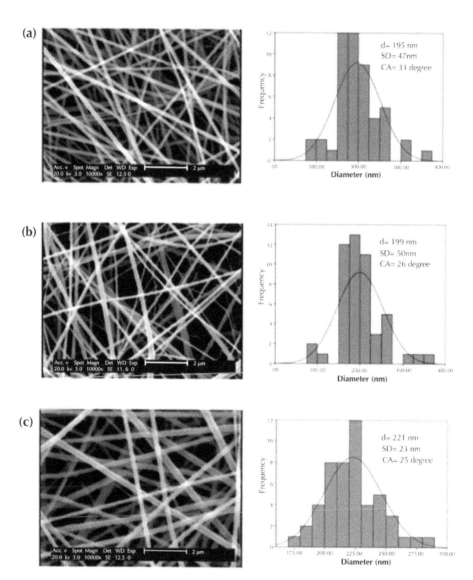

FIGURE 6   The SEM images and fiber diameter distributions of electrospun fibers in volume flow rate of (a) 2 ml/hr, (b) 2.5 ml/hr, and (c) 3 ml/hr.

12.3.2 The Analysis of Variance (ANOVA)

All thirty experimental runs of CCD were performed as described in Table 2. A significance level of 5% was selected: that is statistical conclusions may be assessed with 95% confidence. In this significance level, the factor has significant impact on response if the p-value is less than 0.05. And when p-value is greater than 0.05, it is concluded the factor has no significant effect on response.

TABLE 2   The actual design of experiments and responses for AFD and CA.

No.	Electrospinning parameters				Responses	
	X_1 Concentration	X_2 Voltage	X_3 Distance	X_4 Flow rate	AFD (nm)	CA (°)
1	10	14	10	2	206±33	44±6
2	10	22	10	2	187±50	54±7
3	10	14	20	2	162±25	61±6
4	10	22	20	2	164±51	65±4
5	10	14	10	3	225±41	38±5
6	10	22	10	3	196±53	49±4
7	10	14	20	3	181±43	51±5
8	10	22	20	3	170±50	56±5
9	10	18	15	2.5	188±49	48±3
10	12	14	15	2.5	210±31	30±3
11	12	22	15	2.5	184±47	35±5
12	12	18	10	2.5	214±38	22±3
13	12	18	20	2.5	205±31	30±4
14	12	18	15	2	195±47	33±4
15	12	18	15	3	221±23	25±3
16	12	18	15	2.5	199±50	26±4
17	12	18	15	2.5	205±31	29±3
18	12	18	15	2.5	225±38	28±5
19	12	18	15	2.5	221±23	25±4

TABLE 2 *(Continued)*

| No. | Electrospinning parameters | | | | Responses | |
	X_1 Concentration	X_2 Voltage	X_3 Distance	X_4 Flow rate	AFD (nm)	CA (°)
20	12	18	15	2.5	215±35	24±3
21	12	18	15	2.5	218±30	21±3
22	14	14	10	2	255±38	31±4
23	14	22	10	2	213±37	35±5
24	14	14	20	2	240±33	33±6
25	14	22	20	2	200±30	37±4
26	14	14	10	3	303±36	19±3
27	14	22	10	3	256±40	28±3
28	14	14	20	3	283±48	39±5
29	14	22	20	3	220±41	36±4
30	14	18	15	2.5	270±43	20±3

The results of ANOVA for AFD and CA of electrospun fiber mat are shown in Table 3 and Table 4 respectively. Equations (2) and (3) are the calculated regression equation.

TABLE 3 Analysis of variance for AFD.

Source	Sum of squares	F-value	p-value
Model	31004.72	28.67	< 0.0001
X_1	17484.50	226.34	< 0.0001
X_2	4201.39	54.39	< 0.0001
X_3	2938.89	38.04	< 0.0001
X_4	3016.06	39.04	< 0.0001
X_1X_2	1139.06	14.75	0.0016

TABLE 3   *(Continued)*

Source	Sum of squares	F-value	p-value
X_1X_3	175.56	2.27	0.1524
X_1X_4	637.56	8.25	0.0116
X_2X_3	39.06	0.51	0.4879
X_2X_4	162.56	2.10	0.1675
X_3X_4	60.06	0.78	0.3918
X_1^2	945.71	12.24	0.0032
X_2^2	430.80	5.58	0.0322
X_3^2	0.40	0.005	0.9433
X_4^2	9.30	0.12	0.7334
Lack of Fit	711.41	0.8	Not significant

$$R^2 = 0.9640; R_{adj}^2 = 0.9303.$$

TABLE 4   Analysis of variance for contact angle (CA) of electrospun fiber mat.

Source	Sum of squares	F-value	p-value
Model	4175.07	32.70	< 0.0001
X_1	1760.22	193.01	< 0.0001
X_2	84.50	9.27	0.0082
X_3	338.00	37.06	< 0.0001
X_4	98.00	10.75	0.0051
X_1X_2	42.25	4.63	0.0116
X_1X_3	42.25	4.63	0.0116
X_1X_4	42.25	4.63	0.0116

TABLE 4 *(Continued)*

Source	Sum of squares	F-value	p-value
X_2X_3	12.25	1.34	0.2646
X_2X_4	6.25	0.69	0.4207
X_3X_4	2.25	0.25	0.6266
X_1^2	161.84	17.75	0.0008
X_2^2	106.24	11.65	0.0039
X_3^2	0.024	0.003	0.9597
X_4^2	21.84	2.40	0.1426
Lack of Fit	95.30	1.15	Not significant

$$R^2 = 0.9683; R^2_{adj} = 0.9387.$$

$$
\begin{aligned}
AFD = {} & 212.11 + 31.17X_1 - 15.28X_2 - 12.78X_3 + 12.94X_4 \\
& -8.44X_1X_2 + 3.31X_1X_3 + 6.31X_1X_4 + 1.56X_2X_3 - 3.19X_2X_4 - 1.94X_3X_4 \\
& +19.11X_1^2 - 12.89X_2^2 - 0.39X_3^2 - 1.89X_4^2
\end{aligned}
\tag{2}
$$

$$
\begin{aligned}
CA = {} & 25.80 - 9.89X_1 + 2.17X_2 + 4.33X_3 - 2.33X_4 \\
& -1.63X_1X_2 - 1.63X_1X_3 + 1.63X_1X_4 - 0.88X_2X_3 - 0.63X_2X_4 + 0.37X_3X_4 \\
& +7.90X_1^2 + 6.40X_2^2 - 0.096X_3^2 + 2.90X_4^2
\end{aligned}
\tag{3}
$$

From the p-values presented in Table 3 and Table 4, it can be concluded that the p-values of terms X_3^2, X_4^2, X_2X_3, X_1X_3, X_2X_4, and X_3X_4 in the model of AFD and X_3^2, X_4^2, X_2X_3, X_2X_4, and X_3X_4 in the model of CA is greater than the significance level of 0.05, therefore they have no significant effect on corresponding response. Since the terms had no significant effect on AFD and CA of electrospun fiber mat, these terms were removed and fitted the equations by regression analysis again. The fitted equations in coded unit are given in Equations (4) and (5).

$$
\begin{aligned}
AFD = {} & 211.89 + 31.17X_1 - 15.28X_2 - 12.78X_3 + 12.94X_4 \\
& -8.44X_1X_2 + 6.31X_1X_4 \\
& +18.15X_1^2 - 13.85X_2^2
\end{aligned}
\tag{4}
$$

$$CA = 26.07 - 9.89X_1 + 2.17X_2 + 4.33X_3 - 2.33X_4$$
$$- 1.63X_1X_2 - 1.63X_1X_3 + 1.63X_1X_4 \qquad (5)$$
$$+ 9.08X_1^2 + 7.58X_2^2$$

Now, all the p-values are less than the significance level of 0.05.

The predicted *vs* actual plots for AFD and CA of electrospun fiber mat are shown in Figures 7 and 8 respectively. Actual values are the measured response data for a particular run, and the predicted values evaluated from the model. These plots have determination coefficient (R^2) of 0.9640 and 0.9683 for AFD and CA respectively. It can be observed that experimental values are in good agreement with the predicted values.

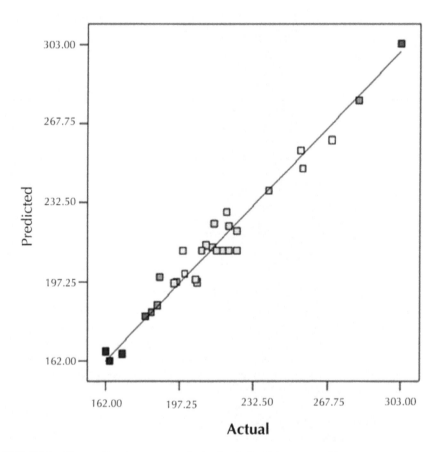

FIGURE 7   The predicted versus actual plot for AFD of electrospun fiber mat.

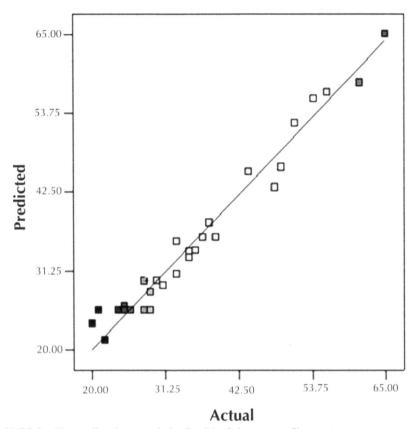

FIGURE 8   The predicted *vs* actual plot for CA of electrospun fiber mat.

12.4 RESPONSE SURFACES FOR AFD

12.4.1 Solution Concentration

Generally, a minimum solution concentration is required to obtain uniform fibers from electrospinning. This concentration, polymer chain entanglements are insufficient and a mixture of beads and fibers is obtained. As the solution concentration increases, the shape of the beads changes from spherical to spindle like [17].

The AFD increased with solution concentration as shown in Figures 9(a), (b), and (c) that is in agreement with observations [28-30]. Figure 9(a) shows the effect of changing solution concentration and applied voltage at fixed spinning distance and volume flow rate. It can be seen the increase in AFD with increase in solution concentration at any given voltage. As shown in Figure 9(b), no interaction was observed between solution concentration and spinning distance. This means that the function of solution concentration was independent from spinning distance for AFD. The effect of solution concentration on AFD was influenced by volume flow rate (Figure 9(c)) and this agrees the presence of the term X_1X_4 in the model of AFD.

12.4.2　Applied Voltage

Figure 9(a), (d), and (e) show the effect of applied voltage on AFD. In this work, the AFD suffer an increase followed by a decrease with increasing the applied voltage. The surface plot in Figure 9(a) indicated that there was a considerable interaction between applied voltage and solution concentration and this is in agreement with the presence of the term X_1X_2 in the model of AFD. In the present study, applied voltage influenced AFD regardless of spinning distance and volume flow rate as shown in Figures 9(d) and (e). The absence of X_2X_3 and X_2X_4 in the model of AFD proves this observation.

12.4.3　Tip to Collector Distance

The tip to collector distance was found to be another important processing parameter as it influences the solvent evaporating rate and deposition time as well as electrostatic field strength. Figures 9(b), (d), and (f) represents the decrease in AFD with spinning distance. It was founded that spinning distance influences AFD independent from solution concentration, applied voltage, and volume flow rate. This agrees the absence of X_1X_3, X_2X_3 and X_3X_4 in the model of AFD.

12.4.4　Volume Flow Rate

The effect of volume flow rate on AFD is shown in Figures 9(c), (e), and (f) indicated the increase in AFD with volume flow rate. Figure 9(c) shows the surface plot of interaction between volume flow rate and solution concentration. It can be seen that at fixed applied voltage and spinning distance, the increase in volume flow rate and solution concentration result the higher AFD. As depicted in Figures 9(e) and (f), the effect of volume flow rate on AFD was independent from applied voltage and spinning distance. This observation confirms the absence of X_2X_4 and X_3X_4 in the model of AFD.

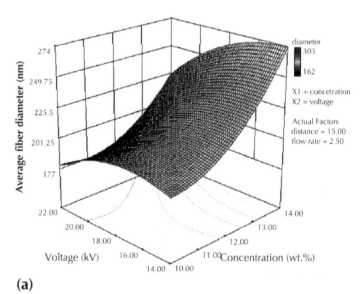

(a)

FIGURE 9　*(Continued)*

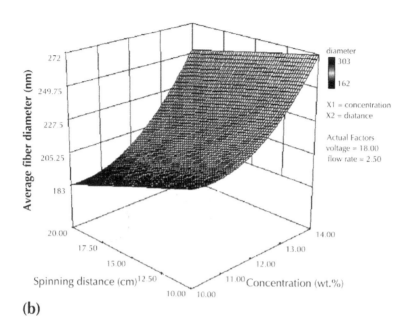

(b)

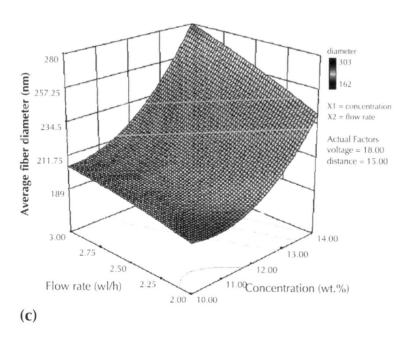

(c)

FIGURE 9   *(Continued)*

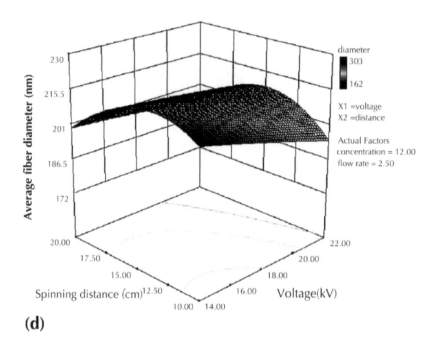

(d)

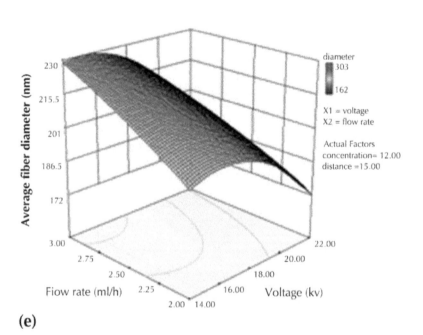

(e)

FIGURE 9 *(Continued)*

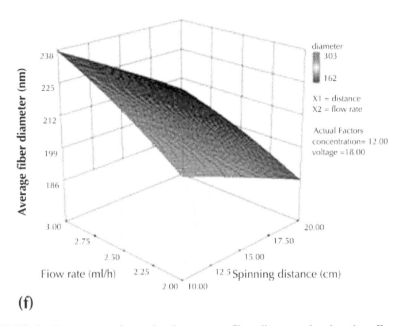

(f)

FIGURE 9   Response surfaces plot for average fiber diameter showing the effect of: (a) solution concentration and applied voltage, (b) solution concentration and spinning distance, (c) solution concentration and volume flow rate, (d) applied voltage and spinning distance, (e) applied voltage and flow rate, and (f) spinning distance and volume flow rate.

12.5   RESPONSE SURFACES FOR CONTACT ANGLE (CA) OF ELECTROSPUN FIBER MAT

12.5.1   Solution Concentration

Figures 10(a), (b), and (c) show the effect of solution concentration on CA of electro-spun fiber mat. In this work, the CA of electrospun fiber mat decrease with increasing the solution concentration.

Figure 10(a) shows the surface plot of interaction between solution concentration and applied voltage. It is obvious that at fixed spinning distance and volume flow rate, the increase in applied voltage and decrease in solution concentration result the higher CA. As shown in Figure 10(b), there was a considerable interaction between solution concentration and spinning distance and this is in agreement with the presence of the term X_1X_3 in the model of CA. The surface plot in Figure 10(c) shows the interaction between solution concentration and volume flow rate at fixed applied voltage and spin-ning distance. It can be seen that at any given flow rate, CA of electrospun fiber mat will increase as solution concentration decreases.

12.5.2   Applied Voltage

The effect of applied voltage on CA of electrospun fiber mat is shown in Figures 10(a), (d), and (e). It can be seen that the CA suffer a decrease followed by an increase with increasing the applied voltage. As depicted in Figure 10(a), the impact of applied voltage on CA of electrospun fiber mat will change at different solution concentration. Figure 10(d) shows that there was no combined effect between applied voltage and spinning distance. Also no interaction was observed between applied voltage and volume flow rate (Figure 10(e)). Therefore, applied voltage had interaction with solution concentration which had been confirmed by the existence of the term X_1X_2 in the model of CA.

12.5.3 Tip to Collector Distance

The impact of spinning distance on CA of electrospun fiber mat is illustrated in Figures 10(b), (d), and (f). Increasing the spinning distance causes the CA of electrospun fiber mat to increase. As demonstrated in Figure 10(b), low solution concentration cause the increase in CA of electrospun fiber mat at large spinning distance. Spinning distance affected CA of electrospun fiber mat regardless of applied voltage and volume flow rate (Figures 10(d) and (f)) as could be concluded from the model of CA. This means that no interaction exists between these variables.

12.5.4 Volume Flow Rate

The surface plot in Figures 10(c), (e), and (f) represented the effect of volume flow rate on CA of electrospun fiber mat. Figure 10(c) shows the interaction between volume flow rate and solution concentration. As illustrated in Figures 10(e) and (f), the effect of volume flow rate on CA of electrospun fiber mat was independent from applied voltage and spinning distance.

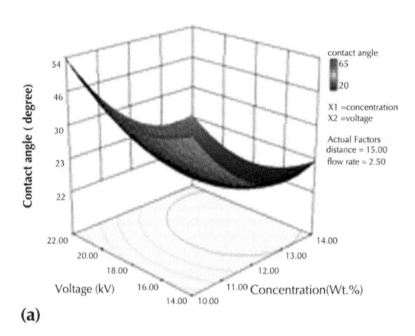

(a)

FIGURE 10   *(Continued)*

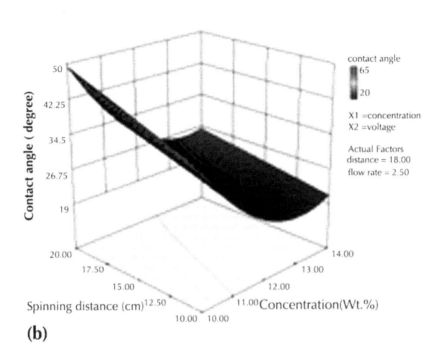

(b)

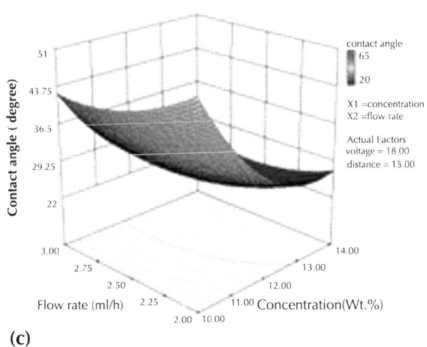

(c)

FIGURE 10 *(Continued)*

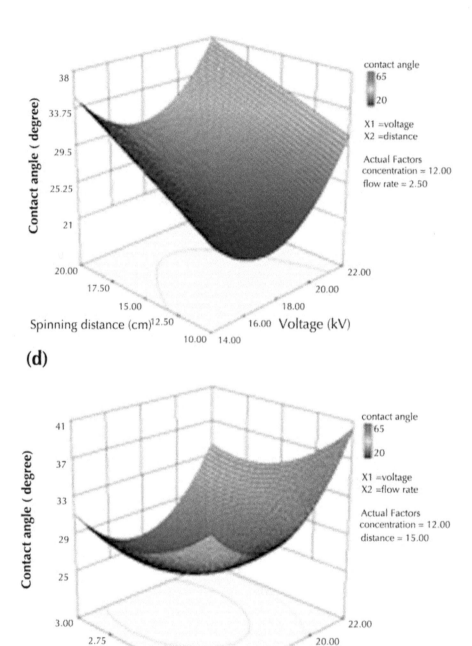

(d)

(e)

FIGURE 10   *(Continued)*

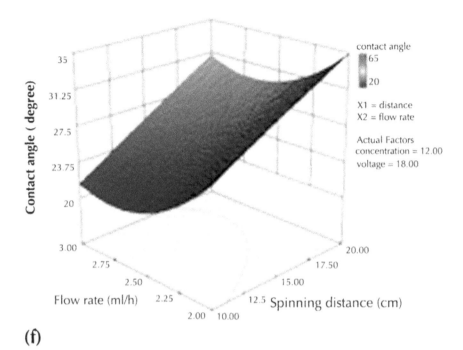

(f)

FIGURE 10 Response surfaces plot for contact angle of electrospun fiber mat showing the effect of: (a) solution concentration and applied voltage, (b) solution concentration and spinning distance, (c) solution concentration and volume flow rate, (d) applied voltage and spinning distance, (e) applied voltage and flow rate, and (f) spinning distance and volume flow rate.

12.6 DETERMINATION OF OPTIMAL CONDITIONS

The optimal conditions were established by desirability. Independent variables namely solution concentration, applied voltage, spinning distance, and volume flow rate were set in range and dependent variable (contact angle) was fixed at minimum. The optimal conditions in the tested range for minimum CA of electrospun fiber mat are shown in Table 5.

TABLE 5 Optimum values of the process parameters for minimum contact angle of electrospun fiber mat.

Parameter	Optimum value
Solution concentration (wt.%)	13.2
Applied voltage (kV)	16.5
Spinning distance (cm)	10.6
Volume flow rate (ml/h)	2.5

12.7   RELATIONSHIP BETWEEN AFD AND CA OF ELECTROSPUN FIBER MAT

The wettability of surface controlled both by the surface chemistry and surface roughness. The morphology and structure of electrospun fiber mat, such as the nanoscale fibers and interfibrillar distance, increases the surface roughness as well as the fraction of contact area of droplet with the air trapped between fibers. It is proved that the CA can provide valuable information about surface roughness. Figure 11 shows the variation of CA with AFD. The CA is observed to decrease with the increase in AFD, which is in good agreement with other report [32]. It can be concluded that the thinner fibers, due to their high surface roughness, have higher CA than the thicker fibers.

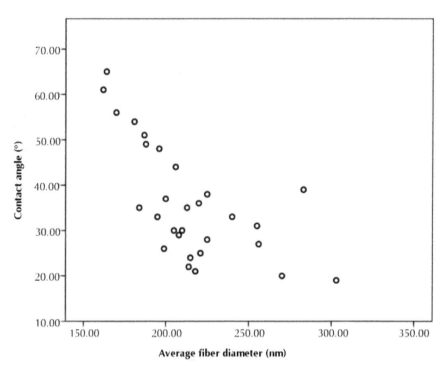

FIGURE 11   Variation of contact angle with average fiber diameter.

12.8   CONCLUSION

In this work, the effects of four electrospinning parameters (solution concentration, applied voltage, tip to collector distance, and volume flow rate) on AFD and CA of PAN nanofibers mat were investigated using RSM. The CCD was used to establish a quantitative relationship between factors and corresponding responses. The results showed that polymer solution concentration was the most significant factor impacting the AFD and CA of electrospun fiber mat. The RSM was successfully applied to find out the optimum level of the factors. The optimum solution concentration, applied voltage, spinning distance, and flow rate were found to be 13.2 wt.%, 16.5 kV, 10.6 cm, and 2.5 ml/hr, respectively, for minimum CA angle of electrospun fiber mat. The values obtained from the model were in good agreement with the experimental values and a good determination coefficient (R^2) of 0.9640 and 0.9683 was obtained for AFD and CA respectively. This study also suggests that the thin fibers exhibit high surface roughness as well as high contact angle.

KEYWORDS

- **Analysis of variance**
- **Average fiber diameter**
- **Central composite design**
- **Contact angle**
- **Response surface methodology**

REFERENCES

1. Shams Nateri, A. and Hasanzadeh, M. *J. Comput. Theor. Nanosci.*, **6**, 1542–1545 (2009).
2. Kilic, A., Oruc, F., and Demir, A. *Text. Res. J.*, **78**, 532–539 (2008).
3. Dotti, F., Varesano, A., Montarsolo, A., Aluigi, A., Tonin, C., and Mazzuchetti, G. *J. Ind. Text.*, **37**, 151–162 (2007).
4. Lu, Y., Jiang, H., Tu, K., and Wang, L. *Acta Biomater.*, **5**, 1562–1574 (2009).
5. Lu, H., Chen, W., Xing, Y., Ying, D., and Jiang, B. *J. Bioact. Compat. Pol.*, **24**, 158–168 (2009).
6. Nisbet, D. R., Forsythe, J. S., Shen, W., Finkelstein, D. I., and Horne, M. K. *J. Biomater. Appl.*, **24**, 7–29 (2009).
7. Ma, Z., Kotaki, M., Inai, R., and Ramakrishna, S. *Tissue Eng.*, **11**, 101–109 (2005).
8. Hong, K. H. *Polym. Eng. Sci.*, **47**, 43–49 (2007).
9. Zhang, W. and Pintauro, P. N. *ChemSusChem.*, **4**, 1753–1757 (2011).
10. Lee, S. and Obendorf, S. K. *Text. Res. J.*, **77**, 696–702 (2007).
11. Reneker, D. H. and Chun, I. *Nanotechnology*, **7**, 216–223 (1996).
12. Shin, Y. M., Hohman, M. M., Brenner, M. P., and Rutledge, G. C. *Polymer*, **42**, 9955–9967 (2001).
13. Reneker, D. H., Yarin, A. L., Fong, H., and Koombhongse, S. *J. Appl. Phys.*, **87**, 4531–4547 (2000).
14. Zhang, S., Shim, W. S., and Kim, J. *Mater. Design*, **30**, 3659–3666 (2009).
15. Yördem, O. S., Papila, M., and Menceloğlu, Y. Z. *Mater. Design*, **29**, 34–44 (2008).
16. Chronakis, I. S. *J. Mater. Process. Tech.*, **167**, 283–293 (2005).
17. Haghi, A. K. and Akbari, M. *Phys. Status. Solidi. A.*, **204**, 1830–1834 (2007).
18. Zhu, M., Zuo, W., Yu, H., Yang, W., and Chen, Y. *J. Mater. Sci.*, **41**, 3793–3797 (2006).
19. Ding, B., Kim, H., Lee, S., Lee, D., and Choi, K. *Fiber Polym.*, **3**, 73–79 (2002).
20. Kanafchian, M., Valizadeh, M., and Haghi, A. K. *Korean J. Chem. Eng.*, **28**, 445–448 (2011).
21. Miwa, M., Nakajima, A., Fujishima, A., Hashimoto, K., and Watanabe, T. *Langmuir*, **16**, 5754–5760 (2000).
22. Öner, D. and McCarthy, T. J. *Langmuir*, **16**, 7777–7782 (2000).
23. Abdelsalam, M. E., Bartlett, P. N., Kelf, T., and Baumberg, J. *Langmuir*, **21**, 1753–1757 (2005).
24. Nakajima, A., Hashimoto, K., Watanabe, T., Takai, K., Yamauchi, G., and Fujishima, A. *Langmuir*, **16**, 7044–7047 (2000).
25. Zhong, W., Liu, S., Chen, X., Wang, Y., and Yang, W. *Macromolecules*, **39**, 3224–3230 (2006).

26. Myers, R. H., Montgomery, D. C., and Anderson-Cook, C. M. *Response surface methodology process and product optimization using designed experiments, 3rd ed.*, John Wiley and Sons, USA (2009).
27. Gu, S. Y., Ren, J., and Vancso, G. J. *Eur. Polym. J.*, **41**, 2559–2568 (2005).
28. Zhang, S., Shim, W. S., and Kim, J. *Mater. Design*, **30**, 3659–3666 (2009).
29. Zhang, C., Yuan, X., Wu, L., Han, Y., and Sheng, J. *Eur. Polym. J.*, **41**, 423–432 (2005).
30. Ziabari, M., Mottaghitalab, V., and Haghi, A. K. In *Nanofibers Fabrication, Performance, and Applications.* W. N. Chang (Ed.). Nova Science Publishers, USA Chapter 4, pp. 153–182 (2009).
31. Ramakrishna, S., Fujihara, K., Teo, W. E., Lim, T. C., and Ma, Z. *An Introduction to Electrospinning and Nanofibers.* National University of Singapore, World Scientific Publishing, Singapore (2005).
32. Ma, M., Mao, Y., Gupta, M., Gleason, K. K., and Rutledge, G. C. *Macromolecules*, **38**, 9742–9748 (2005).

13 Recycled Thermoset Plastics

CONTENTS

13.1 INTRODUCTION

Plastic waste management is one of the major environmental concerns in the world. Plastics can be separated into two types. The first type is thermoplastic, which can be melted for recycling in the plastic industry, such as polyethylene terephthalate (PET). The second type is thermosetting plastic. This plastic cannot be melted by heating such as melamine. At present, these plastic wastes are disposed by either burning or burying. Therefore, both the ways contributing to the environmental problems. This chapter describes the use of thermosetting plastic waste as aggregate within lightweight concrete for building application. The ultimate aim of this chapter was to determine the suitable proportion to achieve the lowest dry density and acceptable compressive strength for non-load-bearing lightweight concrete according to ASTM C129 Type II standard. Experimental tests for the variation of mix proportion were carried out to determine the suitable proportion to achieve the required properties of standard. The mix design in this research is the proportion of plastic, sand, water-cement ratio, aluminum powder, silica fume, and superplasticizer.

The results presented show that with an increase of replacement ratio of materials of lightweight concrete by melamine plastic aggregates the compressive strength and densities of the lightweight concrete were reduced and the absorption and volume of permeable voids of the lightweight concrete were increased. The scanning electron microscopy analysis of composites reveals that cement paste melamine aggregates adhesion is imperfect and weak. Also, the scanning electron microscopy shows that with an increase of replacement ratio of materials of lightweight concrete by melamine plastic aggregates, the pores and cavernous in the structure of the lightweight concrete were increased. It was found that the optimum proportions of materials are cement: aluminum powder: water: sand: melamine: silica fume: superplasticizer equal to 1.0 :0.004:0.35:1.4:2.0:0.25:0.007. The results of compressive strength and dry density are 7.06 N/mm^2 and 887 kg/m^3, respectively. The value of compressive strength and density satisfies the specification for non-load-bearing lightweight concrete according to [1] Type II standard.

Plastic materials production has reached global maximum capacities leveling at 260 million tones in 2007, where in 1990 the global production capacity was estimated at an 80 million tones. It is estimated that production of plastics worldwide is growing at a rate of about 5% per year [2]. Its low density, strength, user friendly designs, fabrication capabilities, long life, light weight, and low cost are the factors behind such phenomenal growth. Plastics also contribute to our daily life functions in many aspects. With such large and varying applications, plastics contribute to an ever increasing volume in the solid waste stream [3].

Each year, Iran produces well into the 100 millions tones of waste, including a sizable percentage of plastic wastes. The major problems that this level of waste production generates initially entail storage and elimination. Storage implies the availability of large surface areas and their management, which in turn requires the installation of specific containers according to the nature of the products that are deposited there. Elimination is most typically carried out by either land filling or incineration (regardless of the type of plastic waste) and by regeneration (only in the case of thermoplastic products) [4, 5]. The burning causes to emissions of carbon dioxide (CO_2), nitrogen oxide (NO), and sulfur dioxide (SO_2). On the other hand, plastic materials remain in the environment for 100, perhaps 1,000 of years. Therefore, both the ways contributing to the environmental problems. Within this framework, the objective of this chapter is to find alternatives for managing the plastic wastes while protecting the environment. To achieve this purpose, a study of these thermosetting plastics has been recycled for use into construction materials.

The self weight of concrete elements is high and can represent a large proportion of the load on a structure. Therefore, using lightweight concrete with a lower density can result in significant benefits such as superior load bearing capacity of elements, smaller cross sections and reduced foundation sizes. A lightweight structure is also desirable in earthquake prone areas [6]. It is convenient to classify the various types of lightweight concrete by their method of production. These are [7]:

(1) By using porous lightweight aggregate of low apparent specific gravity, that is lower than 2.6, for example, pumice material. This type of concrete is known as lightweight aggregate concrete.

(2) By introducing large voids within the concrete or mortar mass: these voids should be clearly distinguished from the extremely fine voids produced by air entrainment such as aluminum powder. This type of concrete is variously known as aerated, cellular, foamed, or gas concrete.

(3) By omitting the fine aggregate from the mix so that a large number of interstitial voids are present: normal weight coarse aggregate is generally used. This concrete is known as no-fines concrete.

In this research, the technology of aerated concrete to produce lightweight concrete has been employed. Since the specific gravity of thermosetting plastic is about one-half of the typical fine and coarse aggregates, therefore, application of thermosetting plastic waste in the concrete corresponds to the first type of lightweight concrete.

The possible use of recycled plastic waste in concrete and other construction materials has been studied by a number of researchers. It was reported by Naik et al. [8], that compressive strength decreased with an increase in the amount of the plastic in concrete, particularly above 0.5% plastic addition to total weight of the mixture. Choi et al. [9] investigated the quality of lightweight aggregates, conducting tests on the workability and the strength properties of concrete, analyzing the relationship between the quality of aggregates and the properties of concrete. Lightweight aggregates were made from waste PET bottles, and granulated blast-furnace slag (GBFS) was used to examine whether it is possible to improve the quality of lightweight aggregate. The 28 days compressive strength of waste PET bottles lightweight aggregate concrete (WPLAC) with the replacement ratio of 75% reduces about 33% compared to the control concrete in the water-cement ratio of 45%. The density of WPLAC varies from 1940 to 2260 kg/m³ by the influence of waste PET bottles lightweight aggregate (WPLA). The structural efficiency of WPLAC decreases as the replacement ratio increases. The workability of concrete with 75% WPLA improves about 123% compared to that of the normal concrete in the water-cement ratio of 53%. The adhered GBFS is able to strengthen the surface of WPLA and to narrow the transition zone owing to the reaction with calcium hydroxide. Marzouk et al. [4] studied the effects of PET waste on the density and compressive strength of concrete. It was found that the density and compressive strength decreased when the PET aggregates exceeded 50% by volume of sand. The density and compressive strength of concrete were between

1,000–2,000 kg/m^3 and 5–60 MPa, respectively. Batayneh et al. [10] investigated the performance of the ordinary Portland cement (OPC) concrete mix under the effect of using recycled waste materials, namely glass, plastics, and crushed concrete as a fraction of the aggregates used in the mix. The main findings of this investigation revealed that the three types of waste materials could be reused successfully as partial substitutes for sand or coarse aggregates in concrete mixtures. Similarly, the results of Ismail and Al-Hashmi [11] showed that reusing waste plastic as a sand substitution aggregate in concrete gives a good approach to reduce the cost of materials and solve some of the solid waste problems posed by plastics. Recently, Albano et al. [12] and Choi et al. [6] also studied the mechanical behavior of concrete with recycled PET, varying the PET content. Results indicate that when volume proportion of PET increased in concrete, showed a decrease in compressive strength, however, the water absorption increased.

The main objective of this study is to investigate the possibility of improvement results of the studied by Panyakapo and Panyakapo [5] that used of melamine as thermosetting plastics for application into construction materials has been conducted, particularly for the concrete wall in buildings. Moreover, the characteristics of absorption and voids have been tested. To study the relationship between mechanical properties and composite structure, scanning electron microscopy technique was employed. The ultimate aim of this chapter was to determine the suitable proportion to achieve the lowest dry density and acceptable compressive strength for non-load-bearing lightweight concrete according to [1] Type II standard.

13.2 MATERIALS

The materials used in present study are as follows:

13.2.1 Ordinary Portland Cement

Type I Portland cement conforming to [13] was used.

13.2.2 Sand

Fine aggregate is taken from natural sand. Therefore, it was used after separating by sieve in accordance with the grading requirement for fine aggregate [14]. The sand aggregates were been saturated surface dry. Therefore, sand aggregates immerse in water at approximately 21°C for 24 hr and removing surface moisture by warm air bopping. Table 1 presents the properties of the sand and its gradation is presented in Figure 1.

TABLE 1 Properties of sand and melamine aggregates.

Properties	Sand	Melamine
density (g/cm^3) [5]	2.60	1.574
bulk density (g/cm^3) [15]	–	0.3–0.6
Water absorption (%)	1.64	7.2
Max size (mm)	4.75	1.77
Min size (mm)	–	0.45

TABLE 1 *(Continued)*

Properties	Sand	Melamine
Sieve 200 (%)	0.24	–
Flammability [15]	non flammable	non flammable
Decomposition [15]	–	at > 280 °C formation of NH$_3$

13.2.3 Thermosetting Plastic

Melamine is a widely used type of thermosetting plastic. Therefore, in the present work has been selected for application in the mixed design of composite. The mechanical and physical properties of melamine are shown in Table 1. The melamine waste was ground with a grinding machine. The ground melamine waste was separated under sieve analysis. The results of cumulative percentage passing from a set of sieves were compared with the grading requirements for fine aggregate according to [14]. Scanning electron micrographs (SEM) of melamine aggregates is shown in Figure 2. Those have irregular shape and rough surface texture. The grain size distributions were then plotted as shown in Figure 1. It was observed that the gradation curve of the combination of sand and plastic aggregates after sieve number 16 meets most of the requirements of [14]. Panyakapo and Panyakapo [5] reported that in the process of producing lightweight concrete, aerated concrete was produced by the formation of gas, which rises to the surface of concrete. Therefore, the very small plastic particles tended to rise to the concrete surface along with the rising gas and the relatively large plastic particles segregated to the bottom part of concrete. To avoid the segregation of fine and coarse particles and protect the gradation curve of the combination of sand and plastic aggregates in limited of the requirements of [14], the appropriate particle sizes were chosen from those retained on sieve numbers 10–40.

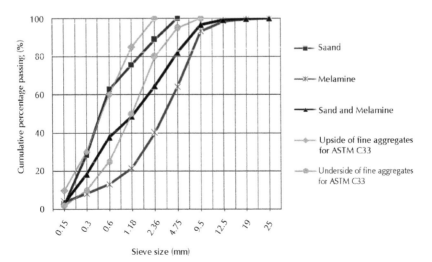

FIGURE 1   Gap-grading analysis of sand and melamine aggregates according to [14].

The melamine aggregates were been saturated surface dry. Therefore, melamine aggregates immerse in water at approximately 21°C for 24 hr and removing surface moisture by warm air bopping.

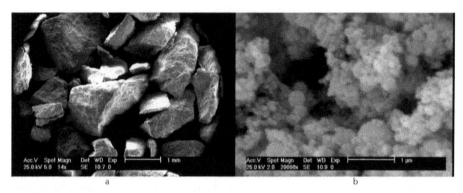

FIGURE 2   The SEM photographs of materials: (a) melamine aggregates and (b) silica fume.

13.2.4   Aluminum Powder

Aluminum powder was selected as an agent to produce hydrogen gas (air entrainment) in the cement. This type of lightweight concrete is then called aerated concrete. The following are possible chemical reactions of aluminum with water:

$$2Al + 6H_2O \longrightarrow 2Al(OH)_3 + 3H_2 \tag{1}$$
$$2Al + 4H_2O \longrightarrow 2AlO(OH) + 3H_2 \tag{2}$$
$$2Al + 3H_2O \longrightarrow Al_2O_3 + 3H \tag{3}$$

The first reaction forms the aluminum hydroxide bayerite ($Al(OH)_3$) and hydrogen, the second reaction forms the aluminum hydroxide boehmite ($AlO(OH)$) and hydrogen, and the third reaction forms aluminum oxide and hydrogen. All these reactions are thermodynamically favorable from room temperature past the melting point of aluminum (660°C). All are also highly exothermic. From room temperature to 280°C, $Al(OH)_3$ is the most stable product, while from 280–480°C, $AlO(OH)$ is most stable. Above 480°C, Al_2O_3 is the most stable product (3) [16, 17]. The following equation illustrates the combined effect of hydrolysis and hydration on tricalcium silicat.

$$CaO.SiO_2 + water \longrightarrow xCaO.ySiO_2(aq.) + Ca(OH)_2 \tag{4}$$

In considering the hydration of Portland cement it is demonstrate that the more basic calcium silicates are hydrolyzed to less basic silicates with the formation of calcium hydroxide or 'slaked lime' as a byproduct. It is this lime which reacts with the aluminum powder to form hydrogen in the making of aerated concrete from Portland cement [18]:

$$2Al + 3Ca(OH)_2 + 6H_2O \longrightarrow 3CaO.Al_2O_3.6H_2O + 3H_2 \tag{5}$$

Hydrogen gas creates many small air (hydrogen gas) bubbles in the cement paste. The density of concrete becomes lower than the normal weight concrete due to this air entrainment.

13.2.5   Silica Fume
In the present work, silica fume has been used. Its chemical compositions and physical properties are being given in Table 2 and Table 3, respectively. The SEM of silica fume is shown in Figure 2.

13.2.6   Superplasticizer
Premia 196 with a density of 1.055 ± 0.010 kg/m^3 was used. It was based on modified polycarboxylate.

TABLE 2   Chemical composition of Silica fume.

Chemical composition	Silica fume
SiO_2 (%)	86–94
Al_2O_3 (%)	0.2–2
Fe_2O_3 (%)	0.2–2.5
C (%)	0.4–1.3
Na_2O (%)	0.2–1.5
K_2O (%)	0.5–3
MgO (%)	0.3–3.5
S (%)	0.1–0.3
CaO (%)	0.1–0.7
Mn (%)	0.1–0.2
SiC (%)	0.1–0.8

TABLE 3   Physical properties of Silica fume.

Items	Silica fume
Specific gravity (gr/cm^3)	2.2–2.3
particle size (μm)	< 1
Specific surface area (m^2/gr)	15–30
Melting point (°C)	1230
Structure	amorphous

13.3   MIX DESIGN

To determine the suitable composition of each material, the mixing proportions were tested in the laboratory, as shown in Table 4. In this study, the mix proportions were separated for five experimental sets. For each set, the cement and aluminum powder contents were specified as a constant proportion. The proportion of each of the remaining materials, that is sand, water, silica fume, aluminum powder, and melamine, was varied for each mix design.

For mix number 1, to determination the primal proportion of melamine plastic content in the composition. The concrete was composed of cement: sand: melamine plastic, adjusted to 1.0:1.0:1.0–3.0. The plastic proportion was increased in increments of 0.5. The water and aluminum powder were taken as 0.35 and 0.004 by weight of cement, respectively.

For mix number 2, to determine the optimum proportion of sand, the quantities of sand by weight of cement were varied from 1.0–1.8 with an increment of 0.2 for each step. The portions of other materials were kept constant. That is, the proportion of cement: aluminum powder: water: melamine plastic were adjusted to 1.0:0.004:0.35:1.0.

For mix number 3, to determine the optimum water content in the composition, the quantities of water were varied from 0.3–0.55 by an increment of 0.05 for each step. The proportions of other materials were kept constant. That is, the composition of cement: aluminum powder: sand: melamine plastic was adjusted to 1.0:0.004:1.4:1.0.

For mix number 4, to determine the optimum proportion of silica fume content in the composition, the quantities of silica fume were varied from 0.1–0.35 by an increment of 0.05 for each step. Due to the reduced workability of the concrete containing silica fume, superplasticizer should be used. It was used to control the slump. Other materials contents were kept the same as the tests. That is, the composition of cement: aluminum powder: sand: water: melamine plastic was adjusted to 1.0:0.004:1.4:0.35:1.0.

For the last mix number, to determine the final proportion of melamine plastic content in the composition, the quantities of melamine plastic were varied from 1.0–2.2 by an increment of 0.2 for each step. Superplasticizer was used to control the slump. The proportions of other materials were kept constant. That is, the composition of cement: aluminum powder: sand: water: silica fume: melamine was adjusted to 1.0:1.4:0.004:0.35:0.25:1.0–2.2.

13.4   EXPERIMENTAL TECHNIQUES

Mortar was mixed in a standard mixer and placed in the standard mold of 50 × 50 × 50 mm according to [19]. In the pouring process of mortar, an expansion of volume due to the aluminum powder reaction had to be considered. The expanded portion of mortar was removed until finishing. The fresh mortar was tested for slump according to [20]. The specimens were cured by wet curing at normal room temperature. The hardened mortar was tested for dry density, compressive strength, water absorption, and voids for the curing age of 7 and 28 days. The test results for melamine, sand, and water contents were reported for 7 days curing age for mix numbers 1–3, because these were very close to the results of 28 days. When silica fume was added in the latter mix numbers 4 and 5, the test results were presented for 28 days. This is because the presence of silica fume increases the duration for completion of the chemical reaction. The testing procedures of dry density, water absorption, and voids were performed according to [21] and compressive strength was performed according to [21].

TABLE 4 Mix proportions of melamine lightweight composites (by weight).

Mix no.	Cement	Aluminum powder	Sand	Water	Silica fume	Melamine	Superplasticizer
1. Determination of melamine content (1st trial mix design)	1.0	0.004	1.0	0.35	–	1.0	–
						1.5	
						2.0	
						2.5	
						3.0	
2. Determination of sand content	1.0	0.004	1.0	0.35	–	1.0	–
			1.2				
			1.4				
			1.6				
			1.8				
3. Determination of water content or water–cement ratio (w/c)	1.0	0.004	1.4	0.30	–	1.0	–
				0.35			
				0.40			
				0.45			
				0.50			
				0.55			
4. Determination of silica fume content	1.0	0.004	1.4	0.35	0.10	1.0	0.005
					0.15		0.007
					0.20		0.009
					0.25		0.012

TABLE 4 *(Continued)*

Mix no.	Cement	Aluminum powder	Sand	Water	Silica fume	Melamine	Superplasticizer
					0.30		0.015
					0.35		0.020
5. Determination of melamine content (final mix design)	1.0	0.004	1.4	0.35	0.25	1.0	0.012
						1.2	0.011
						1.4	0.010
						1.6	0.009
						1.8	0.008
						2.0	0.007
						2.2	0.006

13.5 DISCUSSION AND RESULTS

13.5.1 Mix Number 1 (Determination of Melamine Content for the First Trial Mix Design)

Figure 3 present the variations in the compressive strength and dry density for 7 days age of mortars as a function of the value of melamine substitutes used. It can initially be seen, to increased melamine, the compressive strength and dry density of composites decreased. It was found that the highest compressive strength (3.23 MPa) was obtained for the ratio of cement, aluminum powder, water, sand, and melamine of 1.0:0.004:0.35:1.0:1.0. However, the compressive strength for this mix proportion does not exactly satisfy the standard value. Table 5 shows that the specification of non-load-bearing lightweight concrete according to [1] Type II.

The reduction in the compressive strength due to the addition of melamine aggregates might be due to either a poor bond between the cement paste and the melamine aggregates or to the low strength that is characteristic of plastic aggregates. Due to this reason, the proportion of melamine plastic equal to 1.0 was selected for the next mix design. After the rest of the composition was determined, the melamine plastic was tested again to find the suitable proportion.

TABLE 5 Specification of non-load-bearing lightweight concrete [1].

Type	Compressive strength (MPa)		Density (kg/m3)
	Average of three unit	Individual unit	
II	4.1	3.5	< 1680

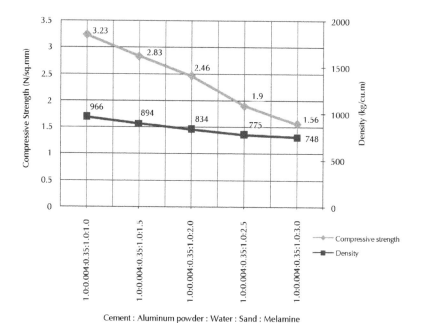

Cement : Aluminum powder : Water : Sand : Melamine

FIGURE 3   Compressive strength and density for varying melamine content (curing for 7 days).

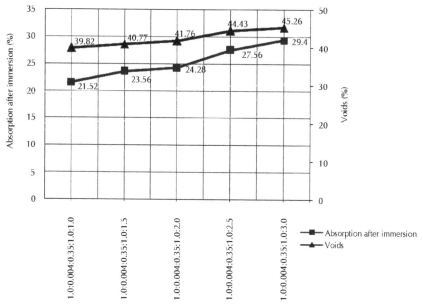

Cement : Aluminum powder : Water : Sand : Melamine

FIGURE 4   Absorption after immersion for varying melamine content (curing for 7 days).

The absorption is an indirect parameter to examine the inside porosity of mortar. The results showed that the absorption after immersion and voids of mortar increased as the melamine content increased (see Figures 4 and 5). Therefore, to increased melamine plastic, the inside porosity of mortar increased. This might be other reason for the reduction in the compressive strength and density.

FIGURE 5   Optical photographs of samples containing varying melamine, right to left containing 1.0, 1.5, 2.0, 2.5, and 3.0 the weight percentage of melamine.

13.5.2   Mix Number 2 (Determination of Sand Content)

The results of compressive strength and dry density for 7 days age are shown in Figure 6. It can be seen that a reduction of sand leads to a reduction in the strength and dry density. The compressive strength and dry density for sand content equal to or greater than 1.4 exactly satisfy the standard value. The proportion of sand equal to 1.4 was selected for the next mix design and the resumed optimize of material. Because, the ultimate aim was to determine the suitable proportion to achieve the lowest dry density and acceptable compressive strength for non-load-bearing lightweight concrete according to [1] Type II standard.

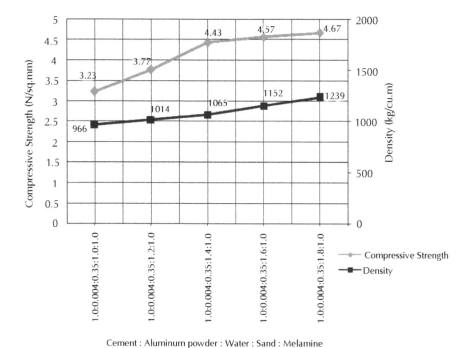

Cement : Aluminum powder : Water : Sand : Melamine

FIGURE 6   Compressive strength and density for the determination of the optimum sand content (curing for 7 days).

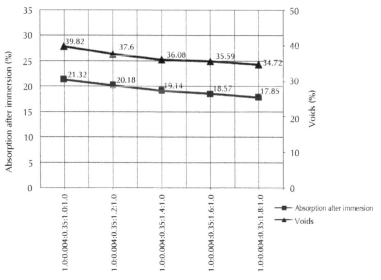

Cement : Aluminum power : Water : Sand : Melamine

FIGURE 7   Absorption after immersion and voids for varying sand content (curing for 7 days).

Figure 7 present the variations in the absorption after immersion and voids as a function of the value of sand substitutes used. The results showed that the absorption after immersion and voids of mortar decreased as the sand content increased. It can be concluded that, the inside porosity of mortar was decreased, when the sand content increased in the mix. The reduction in the absorption after immersion and voids due to the addition of sand might be due to either a good bond between the cement paste and the sand aggregates or to the sand aggregates fine than the melamine aggregates.

13.5.3   Mix Number 3 (Determination of Water Content)

The results of compressive strength and dry density for 7 days age are shown in Figure 8. The results showed that the compressive strength and dry density of mortar decreased as the water content increased.

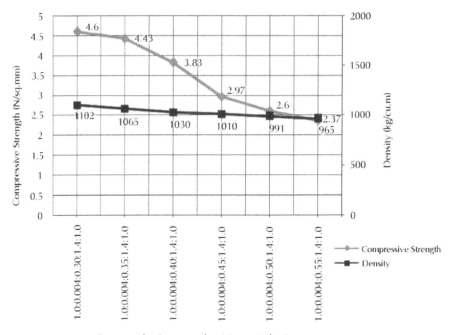

Cement : Aluminum powder : Water : Melamine

FIGURE 8   Compressive strength and density for the determination of the optimum water content (curing for 7 days).

The results of the absorption after immersion and voids tests for varying water content are illustrated in Figure 9. The results showed that the absorption after immersion and voids of mortar increased as the water content increased. It was reported by Choi et al. [9] and Albano et al. [12] that, when the w/c ratios are higher, there is an excess of water that does not participate in the water-cement reaction, so channels with very small diameters like capillaries are produced and when the water evaporates, those empty spaces rest resistance to the concrete. It can be the important reason for decrease of compressive strength and density and increase of absorption after immersion and voids of mortar. It was found that the optimum water content, which leads to the maximum strength, is equal to 0.30. But, to achieve the lowest dry density and acceptable compressive strength, the proportion of water equal to 0.35 was selected for the next mix design.

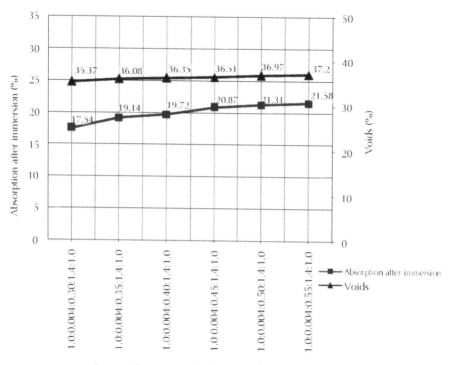

Cement : Aluminum powder : Water : Sand : Melamine

FIGURE 9   Absorption after immersion and voids for varying water content (curing for 7 days).

13.5.4   Mix Number 4 (Determination of Silica Fume Content)

Figure 10 present the variations in the compressive strength and dry density for 28 days age of mortars as a function of the value of silica fume substitutes used. It was found that the results of compressive strength for 7 days age do not increase when compared with those without silica fume. This may be due to the incomplete chemical. The curing age extended to 28 days for the complete chemical reaction between silica fume and water. The figure shows that the compressive strength and dry density of mortar increased as the silica fume content increased to 0.25. But for high silica fume content greater than 0.25, the values of compressive strength decreased. When considering the use of SF as an addition, the micro-filling effect and pozzolanic reaction of SF contributed to a denser microstructure thus resulting in an increase in the compressive strength [22]. A similar result was reported by Rao [23] for the compressive strength of silica fume concrete. An optimum silica fume content of 0.25 was selected as the suitable proportion. The value of compressive strength was equal to 12.06 MPa for silica fume content of 0.25. It was very more than the value of the average of three units (4.1 MPa) as specified for non-load-bearing lightweight concrete according to [1] Type II standard. Therefore, to achieve the lowest dry density and acceptable compressive strength the used more melamine plastic in concrete for the next mix designs.

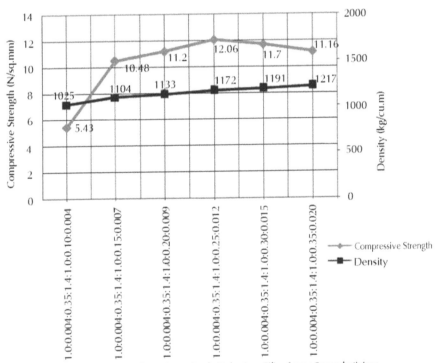

Cement : Aluminum powder : Water : Sand : Melamine : Silica fume : Superplasticizer

FIGURE 10 Compressive strength and density for the determination of the optimum silica fume content (curing for 28 days).

The results of the absorption after immersion and voids tests are shown in Figure 11. It can be seen that the replacement of weight material by silica fume can effectively reduce the absorption and voids. For silica fume concrete, the higher the replacement level, the more the reducing effect on absorption and voids. This reduction is due to the filling effect of SF contributed to a denser microstructure. A similar result was reported by Chan and Ji [24] for the silica fume concrete.

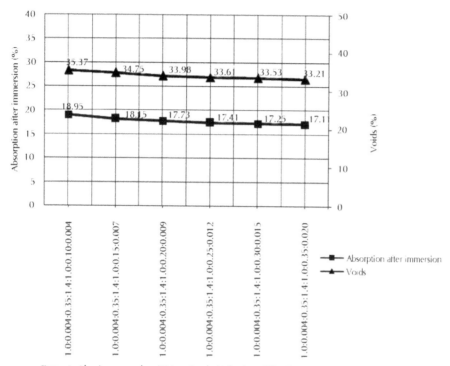

Cement : Aluminum powder : Water : Sand : Melamine : Silica fume : Superplasticizer

FIGURE 11   Absorption after immersion and voids for varying silica fume content (curing for 28 days).

13.5.5 Mix Number 5 (Determination of the Final Melamine Plastic Content)

Figure 12 present the variations in the compressive strength and dry density for 28 days age of mortars as a function of the value of melamine substitutes used. It was found that the presence of melamine caused a reduction in the dry density and compressive strength of concretes as discussed. Figure 13 show that the scanning electron microscopy analysis of composites reveals that cement paste melamine aggregates adhesion is imperfect and weak. Therefore, the problem of bonding between plastic particles and cement paste is main reason to decrease of compressive strength. An optimum melamine content of 2.0 was selected. The results of compressive strength and dry density, which are 7.06 MPa and 887 kg/m^3, are according to [1] Type II standard.

The results showed that the absorption after immersion and voids of mortar increased as the melamine content increased (see Figure 14) as discussed. Also, the structure analysis of mortars by scanning electron microscopy has revealed a low level of compactness in mortars when the value of melamine plastic increased (see Figure 15). It was confirmed that to increased melamine plastic, the inside porosity of mortar increased.

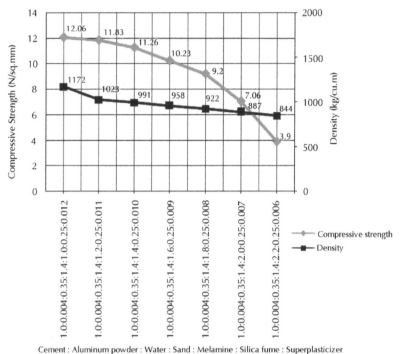

FIGURE 12 Compressive strength and density for the determination of the optimum melamine plastic content (curing for 28 days).

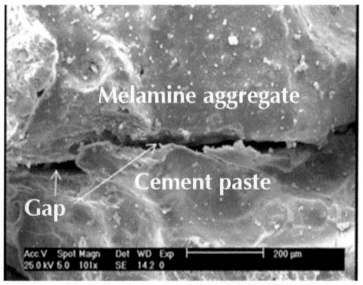

FIGURE 13 Microstructure of concrete containing 1.6 melamine by weight of cement, as obtained using SEM (enlargement: 101×).

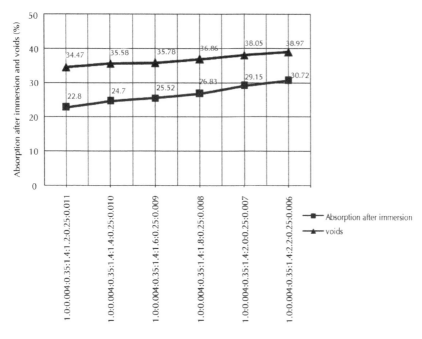

Cement : Aluminum powder : Water : San : Melamine : Silica Fume : Superplasticizer

FIGURE 14   Absorption after immersion and voids for varying melamine plastic content (curing for 28 days).

a

FIGURE 15   *(Continued)*

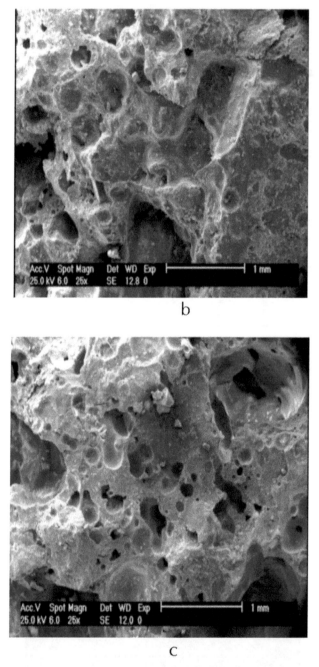

b

c

FIGURE 15 Scanning electron micrographs of various mortars containing melamine plastic aggregates (a) 1.2 by weight of cement (enlargement: 25×), (b) 1.6 by weight of cement (enlargement: 25×), and (c) 2.0 by weight of cement (enlargement: 25×).

13.5.6 Comparison between the Results this chapter and the Results of the Studied by Panyakapo and Panyakapo [5]

Based on the results, the optimum proportions of materials are cement: aluminum powder: water: sand: melamine: silica fume: superplasticizer equal to 1.0:0.004:0.35: 1.4:2.0:0.25:0.007. However, the optimum proportions of materials for the studied by Panyakapo and Panyakapo [5] were cement: sand: water: fly ash: aluminum powder: melamine equal to 1.0:0.8:0.75:0.3:0.0035:0.9. Therefore, the use of melamine plastic more than study (about 122%) for non-load-bearing lightweight concrete according to [1] Type II standard. Also, the dry density of non-load-bearing lightweight concrete was reduction by 36.4% in comparison with the study.

13.5.7 Comparison Research Findings on the Use of Waste Plastic in Concrete

The results of this study are in a perfectly agreement with the other research findings on the use of waste plastic in concrete (see Table 6).

TABLE 6 Comparsion between results of this study and other research findings on the use of waste plastic in concrete (by increase plastic).

Description	Type of waste plastic	Compressive strength	Density	Absorption	void
This study	Melamine	Decrease	Decrease	Increase	Increase
[11]	PET	Decrease	–	Increase	Increase
Kou et al, 2009	PVC	Decrease	Decrease	–	–
[9]	PET	Decrease	Decrease	Increase	Increase
[5]	Melamine	Decrease	Decrease	–	–
[1]	80% polyethylene and 20% polystyrene	Decrease	Decrease	–	–
[5]	PET	Decrease	Decrease	–	–

13.5.8 Comparison between Test Results and Various Standards

The results of compressive strength and dry density are 7.06 N/mm^2 and 887 kg/m^3, respectively. It was found that the compressive strength of plastic lightweight concrete exactly satisfy the class 4 of aerated lightweight concrete according to Institute of Standards and Industrial Research of Iran (ISIRI 8593-1st Edition). However, the dry density is some greater than the ISIRI standard. In addition, the results of this study exactly satisfy the specifications of rendering or plastering mortar for utilization of public and rendering or plastering lightweight mortar according to ISIRI 706-1 standard (1st Revision). Rendering and plastering mortar are widely used in Iran. These comparisons are summarized in Table 7.

TABLE 7 Comparison between the test results of this study and various standards.

Description	Compressive strength (N/mm²) (Average of three units)	Dry density (kg/cm³)
Plastic lightweight concrete (this study), cement : aluminum powder : water : sand : melamine : silica fume : superplasticizer = 1.0 :0.004:0.35:1.4:2.0:0.25:0.007	7.06	887
Aerated concrete masonry units, Class 4 (ISIRI 8593-1st Edition)	5.0	450 – 860
Rendering or plastering mortar for utilization of public (ISIRI 706-1,1st Revision)	0.4 <	–
Rendering or plastering lightweight mortar (ISIRI 706-1,1st Revision)	0.4 – 7.5	≤ 1300

13.6 CONCLUSION

Melamine plastic aggregates can be successfully and effectively utilized for non-load-bearing lightweight concrete according to [1] Type II standard. The following conclusions were drawn from the investigation:

(1) With an increase of replacement ratio of materials of lightweight concrete by melamine plastic aggregates:

(a) The compressive strength and densities of the lightweight concrete were reduced. The reduction in the compressive strength due to the addition of melamine aggregates might be due to either a poor bond between the cement paste and the melamine aggregates or to the low strength that is characteristic of plastic aggregates. Also, to increased melamine plastic, the inside porosity of mortar increased. This might be other reason for the reduction in the compressive strength and density. The other reason for the reduction in the density is due to low specific gravity of melamine plastic than sand.

(b) The absorption and volume of permeable voids of the lightweight concrete were increased.

(2) The existed straight relationship between the compressive strength and density. Also, the existed inverse relationship between the compressive strength, density, absorption, and volume of permeable voids.

(3) Density and compressive strength of mortar decreased as the water content increased. Because, when the w/c ratios are higher, there is an excess of water that does not participate in the water-cement reaction, so channels with very

small diameters like capillaries are produced and when the water evaporates, those empty spaces rest resistance to the concrete.

(4) The compressive strength increase as the addition of SF increases. However, the dry density tends to also increase due to the relatively high specific gravity. Also, absorption and volume of permeable voids decrease as the addition of SF increases. These happen due to the micro-filling effect and pozzolanic reaction of SF contributed to a denser microstructure.

(5) The scanning electron microscopy analysis of composites reveals that cement paste-melamine aggregates adhesion is imperfect and weak. Also, the scanning electron microscopy shows that with an increase of replacement ratio of materials of lightweight concrete by melamine plastic aggregates, the pores and cavernous in the structure of the lightweight concrete were increased.

Utilization of other plastics in the mix proportion for non-load-bearing lightweight concrete according to [1] Type II standard and other standards are suggested for further studies.

KEYWORDS

- **Granulated blast-furnace slag**
- **Melamine**
- **Polyethylene terephthalate**
- **Scanning electron micrographs**
- **Thermoset plastics**

REFERENCES

1. *Annual Book of ASTM Standard*. Standard Specification for Nonloadbearing Concrete Masonry Units, ASTM C129 (2005).
2. Al-Salem, S. M., Lettieri, P., and Baeyens, J. The valorization of plastic solid waste (PSW) by primary to quaternary routes: From re-use to energy and chemicals. *Progress in Energy and Combustion Science*, **36**, 103–129 (2010).
3. Siddique, R., Khatib, J., and Kaur, I. Use of recycled plastic in concrete a review. *Waste Management*, **28**, 1835–1852 (2008).
4. Marzouk, O. Y., Dheilly, R. M., and Queneudec, M. Valorization of post-consumer waste plastic in cementitious concrete composites. *Waste Management*, **27**, 310–318 (2007).
5. Panyakapo, P. and Panyakapo M., Reuse of thermosetting plastic waste for lightweight concrete. *Waste Management*, **28**, 1581–1588 (2008).
6. Choi, Y. W., Moon, D. J., Kim, Y. J., and Lachemi, M. Characteristics of mortar and concrete containing fine aggregate manufactured from recycled waste polyethylene terephthalate bottles. *Construction and Building Materials*, **23**, 2829–2835 (2009).
7. Neville, A. M. and Brooks, J. J. *Concrete Technology*. Longman Scientific and Technology, co-published with John Wiley and Sons, US, p. 345 (1991).
8. Naik, T. R., Singh, S. S., Huber, C. O., and Brodersen, B. S. Use of postconsumer waste plastics in cement-based composites. *Cement and Concrete Research*, **26**(10), 1489–1492 (1996).
9. Choi, Y. W., Moon, D. J., Chung, J. S., and Cho, S. K. Effects of waste PET bottles aggregate on the properties of concrete. *Cement and Concrete Research*, **35**, 776–781 (2005).

10. Batayneh, M., Marie, I., and Asi, I. Use of selected waste materials in concrete mixes. *Waste Management*, **27**, 1870–1876 (2007).

11. Ismail, Z. Z. and Al-Hashmi, E. A. Use of waste plastic in concrete mixture as aggregate replacement. *Waste Management*, **28**, 2041–2047 (2008).

12. Albano, C., Camacho, N., Hernandez, M., Matheus, A., and Gutierrez, A. Influence of content and particle size of waste pet bottles on concrete behavior at different w/c ratios. *Waste Management*, **29**, 2707–2716 (2009).

13. *Annual Book of ASTM Standard*. Specification for Portland Cement, ASTM C150 (1994).

14. *Annual Book of ASTM Standard*. Specification for Fine and Coarse Aggregates, ASTM C33 (1992).

15. Melamine, CAS N:108-78-1. UNEP Publications.

16. U.S. Department of Energy, Reaction of Aluminum with Water to Produce Hydrogen. Version 1.0, Page 1 of 26 (2008).

17. Studart, A. R., Innocentini, M. D. M., Oliveira, I. R., and Pandolfelli, V. C. Reaction of aluminum powder with water in cement-containing refractory castables. *Journal of the European Ceramic Society*, **25**, 3135–3143 (2005).

18. Short, A. and Kinniburgh, W. *Lightweight Concrete*. Applied Science publishers Ltd. London, Third Edition, pp. 17–18 (1978).

19. *Annual Book of ASTM Standard*. Standard Test Method for Compressive Strength of Hydraulic Cement Mortars, ASTM C109 (2002).

20. *Annual Book of ASTM Standard*. Test Method for Slump of Hydraulic–Cement Concrete, ASTM C143 (2003).

21. *Annual Book of ASTM Standard*. Test Method for Density, Absorption, and Voids in Hardened Concrete, ASTM C642 (1997).

22. Wongkeo, W. and Chaipanich, A. Compressive strength, microstructure and thermal analysis of autoclaved and air cured structural lightweight concrete made with coal bottom ash and silica fume. *Materials Science and Engineering*, **30**, 30 (2010).

23. Rao, G. A. Investigations on the performance of silica fume incorporated cement pastes and mortars. *Cement and Concrete Research*, **33**, 1765–1770 (2003).

24. Chan, Y. N. S and Ji, X. Comparative study of the initial surface absorption and chloride diffusion of high performance zeolite, silica fume and PFA concretes. *Cement and Concrete Composites*, **21**, 293–300 (1999).

Index